Biotechnology, Biosafety, and Biodiversity:

Scientific and Ethical Issues for Sustainable Development

editors:
Sivramiah Shantharam and Jane F. Montgomery
Animal and Plant Health Inspection Service, USDA,
Riverdale, Maryland, USA

Science Publishers, Inc.
USA

Library of Congress Cataloging-in-Publication Data

Biotechnology, biosafety, and biodiversity: scientific and ethical issues for sustainable development/Sivramiah Shantharam, Jane F. Montgomery, editors.

p. cm.

Proceedings of the Satellite Symposium on Biotechnology and Biodiversity, held on November 15 and 16, 1996, at the Indian Agricultural Research Institute (IARI), New Delhi, India.

Includes bibliographical references.

ISBN 1-57808-018-5

1. Agricultural biotechnology Congresses. 2. Agrobiodiversity—Congresses. I. Shantharam, Sivramiah, 1951-. II. Montgomery, Jane F. III. Satellite Symposium on Biotechnology and Biodiversity (1996: New Delhi, India)

S494.5.B563B545 1999

631.5'233—dc21 99-14148

CIP

SCIENCE PUBLISHERS, Inc.
Post Office Box 699
Enfield, New Hampshire 03748
United States of America

Internet site: *http://www.scipub.net*

sales@scipub.net (marketing department)
editor@scipub.net (editorial department)
info@scipub.net (for all other enquiries)

This volume may be cited as —
Shantharam, S. and Montgomery, J.F. 1999. Biotechnology and Biodiverisity: Scientific and Ethical Issues for Sustainable Development. Science Publishers, Inc., Enfield, New Hampshire (USA)

ISBN 1-57808-018-5

Published by Science Publishers, Inc., USA
Printed in India

This publication was made possible in part by a grant from the United States Department of Agriculture's Animal and Plant Health Inspection Service (APHIS) and does not necessarily express the views of APHIS or any other U.S. Federal agency.

Foreword

I FEEL PRIVILEGED to be associated with this meeting to discuss the scientific and ethical issues relating to biodiversity and biotechnology. The topic is undeniably important and relevant. It also is admirably complementary to the thrust of the program of the Second International Crop Science Congress that will follow in a couple of days. I would like to congratulate the organizers of this meeting, and especially the moving spirit behind this endeavor, Dr. Shantharam, for their thoughtful initiative.

Science aims at reaching the truth behind what causes an effect. The boundaries of knowledge, extended during the process of establishing cause-and-effect relationship of the phenomenon under investigation, have in many cases led to solutions of problems not apparently related to the initial investigation. Generations of knowledge and the long-term benefits that accrue from it have been advanced by researchers as justification for investments in the pursuit of science. The public, and in recent years even some socially conscious scientists, have started voicing different expectations from science. They believe that investment in science should result in tangible results which improve the socioeconomic status and quality of life of the common person in society. This raises issues, not only of science, per se, but also of how it should be planned and executed. Moral, ethical, and societal issues are increasingly getting on the center stage and this is dramatically true of issues related to biotechnology and biodiversity. The reason for this is that biodiversity and biotechnology are now accepted as important for crop and farm productivity in the future. From a purely scientific viewpoint there is no conflict between biotechnology and biodiversity preservation since biotechnology enables optimal exploitation of biodiversity. On a practical level there are serious concerns, and even doubts, about the benefits of biotechnology for those who need it most. Strength in biotechnology practice and extent of biodiversity do not coexist in a geographical sense. Biodiversity is rich in tropical regions and nations that are economically impoverished. On the other hand, biotechnology has largely been developed in the private sector of biodiversity—poor but economically rich countries.

There are a number of issues that need addressing in the biotechnology-biodiversity debate. I will refer only to some of them, like the impact of scientific innovations in agriculture, biosafety and environmental impacts, socioeconomic implications, and ethical aspects.

I will begin with the question agitating scientists and science managers about the impending shift that the future may see in the conduct of agricultural research. Until now agricultural research was mostly government funded, was done for the public good and had easy and free access to biodiversity available worldwide. The results and products of this research were shared without charge with other researchers, and in some cases even with the for-profit private sector. Intellectual property protection of biotechnology products has now introduced a threatening change. Biotechnology research for product development is done primarily in the private sector in the developed countries. The results and products of private research are available only on payment. This can adversely affect the production capacities of many poor developing countries who in the past have benefitted from free access to improved germplasm, varieties, and agricultural technologies. For example, if the research that developed exploitable dwarfing genes of wheat and rice had been done in the private sector, the developing world would have been denied the fruits of the green revolution.

Also, the research agendas or private companies are, for justifiable reasons, dictated by profit motive. Consequently, only those crops and aspects are pursued where large markets and profits are anticipated. Regional crops and local problems remain untouched by biotechnology firms. Today, when we are interested not only in increasing food productivity and production but also in regional and local self-reliance, we must think of ways by which developing countries can take advantage of technologies that address local crops and problems of the poor farmers.

Questions also are being raised about biosafety of genetically modified organisms, of their field releases, and the consequences of such releases on the environment. Can we extrapolate results of small-scale trials under controlled conditions, which is the method currently used for large-scale commercial cultivation? Can we really predict the impact on wild relatives and nontarget species, especially in the light of interrelatedness of species in natural habitats? The biosafety of engineered food also needs to be examined in a much broader perspective. The food processing and consumption patterns of different communities also deserve consideration. Unequivocal answers to these and many other questions are required to allay public concerns. Also, mechanisms to prevent or limit damage in case of undesirable negative impact have to be thought out.

The socioeconomic impact of biotechnology also deserves careful consideration. For example, who will benefit from the new technology? Will the employment pattern change? How will biotechnology affect farming systems or sustainable agriculture? The questions become relevant because in the short period since the advent of biotechnology in the United States, major changes have been witnessed in both farm production and the agribusiness scenario. Mergers and takeovers of agrichemical and seed companies have been rampant. Monopolistic control regimes are clearly on the increase. We need to ponder if it is a preview of things to come in the 21st Century. Although many biotech innovations appear neutral, the major beneficiaries are the large, scientifically managed farms. Once again, small and marginal farmers of Third World countries may be left out of the biotechnology bonanza, thereby widening the gap between the rich and the poor. Again, presently for the rural unskilled labor, especially women, the major avenue of employment is in agriculture in most Third World countries. If biotechnology promotes large, scientifically managed, intensive farming, the poor, unskilled labor may have to face unemployment. We need to work out ways to ensure that biotechnology does not lead to social inequity.

Product substitution and novel processes to synthesize traditional products are now possible through biotechnology. For example, it is now possible to produce lauric acid from transgenic canola or obtain vanilla in cell cultures. What will be the consequences of such developments on the economy of countries that depend on the export of such commodities? It also is feared that crop diversity and agrobiodiversity may shrink from excessive dependence on a few crop species for obtaining the required products. Should we prevent some of these changes brought about by product substitution and relocation of production centers and trade? Crop production and commodity trade in most countries are subsidized and protected. Therefore, nations will have to prepare suitable alternatives and mutual agreements to tide over such changes.

Legislation is yet another area that demands technical input. Legislation is required to protect innovations and to ensure biosafety. In the absence of suitable legislation, innovators will tend to find protection through secrecy which negates diffusion and wider use of innovative processes and products. Also, unscrupulous individuals or entrepreneurs may take advantage of poor legislation and economic compulsions of developing countries for testing their products. (For example rBST (recombinant Bovine Somatotropin) was approved for use to boost milk production in Mexico in 1990—four years ahead of the U.S. where research and production of rBST originated). How can we protect the interests of illiterate, yet knowledgeable tribal/people whose information is proving vital in bioprocessing? Furthermore, how do we ensure compliance with legislation when research itself is shrouded in secrecy?

Consumer and public education about biotechnology is extremely important to avoid exploitation by vested interest groups. In recent years we have seen severe negative public responses to the introduction of some biotechnology products. If developed countries, that have no compulsion to increase food production, shun or ban use of biotechnology products, Third World countries may also turn skeptical. Greater transparency of experiments and findings is needed to build public confidence in biotechnology.

There are a number of issues to be tackled and many diverse actors in the play. Unless we exchange views and analyze the causes of apprehensions, it is difficult to avoid conflicts. Economic and moral aspects should be given due consideration because, in the words of Dr. de Oliveir, "The laws of the market without any moral and ethical counterpoint are very similar to the laws of the jungle. In the population resource debate it is not Malthus that I fear, but Darwin." The bottom line is that development should not be at a social or environmental cost. I am happy that representatives from various interest groups have assembled here to discuss these issues. It is my hope that the deliberations will help to evolve a consensus approach to exploit biotechnology for the benefit of all.

V. L. Chopra
National Research Centre on Plant Biotechnology
Indian Agricultural Research Institute
New Delhi
India

Preface

WE ARE HAPPY to place this book before you after a regrettable delay of almost two years. The present book embodies the proceedings of the *Satellite Symposium on Biotechnology and Biodiversity: Scientific and Ethical Issues* held on November 15 and 16, 1996, at the Indian Agricultural Research Institute (IARI), New Delhi, India in conjunction with the Second International Crop Science Congress (ICSC).

Agricultural biotechnology is highly advanced and developing at the speed of light. Biotechnology is poised to change the field of agriculture in ever-so-many unprecedented ways and will, no doubt, influence the way we live. Products of agricultural biotechnology are now being commercialized and, in particular, transgenic crops plants are now being planted to millions of acres in the Americas. Transgenic commodity crops are now being traded in the international markets. Some of the major categories of transgenic crop plants are endowed with genetically engineered traits are insect resistance, herbicide tolerance, and virus resistance varieties. With the advent of transgenic crops, people started realizing how biotechnology would change agriculture as we know it, affecting our lifestyles and having major socioeconomic and cultural impacts. In addition, the primary concerns had to do with the biosafety issues and potential environmental impact issues.

Series of debates in the early and mid-1980s resulted in the development of biosafety guidelines and regulatory policies that was designed to facilitate safe transfer of genetically engineered organisms into the field (environment). Concomitant with biotechnology regulatory policies, acute concern was expressed regarding the issue of obtaining intellectual property rights and patent protection for genetically engineered organisms. The debate is raging even to this day is centered around the fundamental issue of patenting living organisms or organisms that are genetically improved using either conventional breeding techniques or biotechnology.

The basic argument has to do with the ethical basis of modern biotechnologists trying to patent genes and transgenic organisms that had been developed up to now through contributions of countless breeders and conserves over generations including communities. Particularly in developing countries where subsistence farmers are so poor that they will not be able to afford the costs of the new and glittering transgenic crop plants. When most of these subsistence farmers have been cultivating their own varieties that had been passed on to them from their ancestors without cost, now they will have to buy it from biotechnology seed companies at a cost that they can ill afford. This point raises another important ethical question of how to compensate farmers, breeders, conserves, communities, or societies for the germplasm used develop genetically engineered crop plants. There have been hundreds of meetings and conferences at the national and international levels to address these fundamental ethical questions to come up with satisfactory answers. The answers are not easy to come by. These debates have become emotionally charged as it hits at the "food" we eat, and food has been so central to all the cultures and civilizations and more so in the modern economy.

This satellite symposium was designed to address some of the key questions related to biosafety, environmental impact issues, and ethical issues affected by the advent of biotechnology in agriculture. Specially invited speakers addressed these questions in their formal presentations and have been included in this book. The speakers were selected on the basis of their intrinsic interest in the subject matter and for their original contributions to the debate in the field who articulated their views in an impartial and professional manner. Their presentations aroused considerable amount of debate in the audience and opinions expressed by all the participants made it a lively two-day session. We ask forgiveness of the participants and the readers for the delay in bringing out this publication. We can assure you that the points raised and covered in this book is current as it has been all these years and they continue to be so for some time to come.

Organizing an international symposium of any magnitude is a daunting task and could not have been done by any one person. First, our thanks are due to the Program Committee of the Second International Crop Science Congress for selecting our proposal to organize the symposium. We would like to thank the United States Department of Agriculture (USDA), Animal and Plant Health Inspection Service (APHIS), for providing generous financial support, and

Dr. John H. Payne, the then Director of Scientific Services, PPQ, for all his encouragement and support. In addition, I would like to mention and to thank Dr. Patrick Rudelsheim of Plant Genetic Systems, Gent, Belgium for his financial contribution to offset some of the costs of the symposium. We would like to express our grateful thanks to Professor Ram Badan Singh, Director of IARI, New Delhi, for providing physical facilities to conduct the symposium; to Dr. R. P. Sharma, Project Director, National Center for Plant Biotechnology, IARI, New Delhi for his efforts in helping local coordination activities; to Dr. Kailash Bansal, Senior Scientist and his students at IARI for their untiring efforts in making all the local arrangements for the symposium. We would like to thank Dr. Anil Kumar Gupta, Center for Agricultural Management, Indian Institute of Management, Ahmedabad for working with us as a co-convener of the symposium. We are indeed grateful to all the speakers and the participants for their contributions and for their enormous patience in waiting for this publication to see the light of day.

Lauren Jones of USDA, APHIS has been singularly responsible for design and coordination of the publication, and our words fail to express our sincere gratitude for her professional efforts. We would like to mention similar thanks to our publishers Oxford & IBH Book House, New Delhi, for their patience and assistance in publishing these proceedings. As usual, they have done an excellent job of printing the book in an elegant fashion.

Please note that although USDA, APHIS provided financial support to this symposium, the opinions expressed in the proceedings of the symposium does not represent the official policy of the position of the agency. All the opinions expressed in the book strictly belong to the authors and individuals who participated in the workshop.

September 1998
Riverdale, Maryland

Sivramiah Shantharam
Jane Montgomery
EDITORS

Contents

Biotechnology and Development: Moving Technology Forward Using Environmental Assessment

Sivramiah Shantharam
U.S. Department of Agriculture, Animal and Plant Health Inspection Service, Riverdale, Maryland, USA

BIOTECHNOLOGY AND BIODIVERSITY are the two important intertwined terms that occupy the center stage of international debate surrounding the public acceptance of biotechnology. Biosafety and environmental impacts of genetically engineered organisms with special reference to biodiversity have created numerous controversies. Environmental impacts of the release of genetically engineered organisms is of critical concern to the scientists and environmentalists alike. More important are the ethical dilemmas related to access to biodviversity, conservation, and reward for the conservation. These concerns take on added measure of complexity when several products in agricultural biotechnology, particularly food and fiber crops, are beginning to be commercialized around the world.

As a priori, field testing of genetically engineered crop plants and drugs will have to undergo careful testing in the environments in which they are going to be cultivated to assess their field level performance and efficacy. While in the North, thousands of field trials of transgenic crop plants have taken place in the past decade, very few field tests have taken place in the developing countries where some of these products are likely to have the greatest impact on agriculture and development. The two major reasons for the lack of progress toward field testing in the developing countries is the lack of skilled scientific and technical professionals to conduct environmental risk assessment and the opposition from environmental activists who demand complete assessment of all aspects of biosafety and environmental risks combined with ethical questions related to intellectual property rights (IPR) issues and patenting of life forms, whether genetically modified or unmodified. Some of the environmental activists are simply opposed to biotechnology per se, which in their opinion allows scientists to play God and threaten the living habitats. Biodiversity is at its center. Absence or ineffective biotechnology regulatory policies in most of the developing countries is a drawback. Clearly there is a need for an impartial body in most of these countries to play the role of a trusted intermediary to assure the concerned public that carefully controlled and closely monitored field tests with genetically engineered organisms is the only way to assess the potential impacts on the environment.

Another important biosafety concern is that the developing countries are a rich source of biodiversity and centrally located for several major crop plants that have been engineered with several important agronomic traits such as insect resistance, herbicide tolerance, and virus resistance. Although a majority of the crops have been engineered with the agronomic traits mentioned, there are many other types of quality related traits that have been engineered into vegetable crops.

Four fundamentals questions surrounding the biosafety and environmental impacts of genetically engineered organisms keep coming up in all environmental assessments. They are as follows:

- Is the introduced gene stably inherited in the transgenic plants and its offsprings?
- What are the routes by which the introduced genes might escape from the engineered plants?
- What are the consequences of such a gene escape?
- Do genetically engineered organisms have the potential to become weeds and aggressive colonizers?

These are the fundamental questions that need to be addressed in any Good Developmental Practice (GDP) for product development.

For the most part, answers to all these questions are available from looking at the history of cultivation of the most important crops like corn, soybean, rice, cotton, and wheat. The most important point in assessing these questions for an environmental risk assessment is to pay close attention to the nature of the genetic engineering done to the plant and the environment into which it is being introduced. In most instances, if sufficient attention is paid to the biological, physical, and temporal safeguards built into the test, one can easily allow controlled and closely monitored field tests to assess the field level performance of the engineered organism, or the plant in this instance. These are the most straight forward biosafety questions which most of the scientific and regulatory scientists can easily defend.

Ticklish questions relate to the long-term effects of growing these plants in monoculture, the associated changes that might occur in agricultural practices, and the socioeconomic benefits of adopting genetically engineered crop plants for large-scale cultivation. Another significant aspect of this debate is the impact of genetically engineered organisms on biodiversity. It should be borne in mind that at the present time there are no useful working models that are quantitative in nature with any sense of predictive value to base the environment risk assessments on. On the other hand, one can estimate the potential impacts or consequences on the environment by using qualitative questions aand by using the knowledge of the reproductive nature of the organism in question. This process has been used successfully at the U.S. Department of Agriculture (USDA), Animal and Plant Health Inspection Service (APHIS) to allow for thousands of field tests since 1987. More importantly, USDA, APHIS was able to gain considerable amount of experience in this area, thereby establishing global leadership in regulating biotechnology. This has allowed for the safe deployment of genetically engineered organisms and has lead to the successful commercialization of more than twenty varieties of genetically engineered crop plants in the United States. It is these years of experience that has allowed USDA, APHIS to pare down the regulatory oversight by introducing a notification system for almost all crops. There is hardly any crop that today needs USDA, APHIS permits for field testing, and those that have been regulated do not come under the purview of U.S. Department of Agriculture (USDA) regulatory authority.

It is clear that the environments in which agriculture is carried out in most tropical countries are of a different nature, and many of them are the centers of origin of the plants in question. It is important to realize this, to seek answers to the same basic questions listed above for environmental risk assessment and to evaluate the kinds of safeguards one needs to put in place for the safe deployment of genetically engineered crop plants.

Biodiversity is under threat in almost all parts of the world, mostly due to human activity. Modern agriculture has contributed its share, and genetically engineered plants will cause additional concern. Monoculture agriculture has allowed domesticating only a select group of about five food crops, and in equal measure the monoculture of millions of hectares of land is being cultivated by just a few varieties of crop plants, purely because of their yield potential. One of the most important reasons why biodiversity is being destroyed is the growth of the human population. Consequently, forests have been cut down. Genetically engineered crop plants are expected to be cultivated in the same areas as it is being cultivated with conventionally bred varieties of crop plants. Large-scale cultivation of genetically engineered crop plants are not expected to escape from well tended cultivation plots and take over the countryside and overrun the wilderness of biodiversity.

There has been no scientific reason or evidence for the aggressive colonization of genetically engineered plants. Gene escape seems to be possible and has been demonstrated in the case of crop plants whose wild and weedy relatives are in the surrounding areas of cultivation. But, whether the escaped gene introgesses into the wild and weedy relatives and whether there will be sufficient selection pressure for the selection and propagation of such genes in the wild, remains an enigmatic question. All the evidence we have from conventionally bred plants for similar traits have not been recorded to have spread in the wild. Only a long-term monitoring of freely cultivated genetically engineered plants (e.g., ten or more years) might shed some light on this question. This exercise turns out to be an academic research problem in plant ecology–one that should not prevent field testing and carefully monitored commercialization of the genetically engineered organism.✧

Biosafety Training Programs and Their Importance in Capacity Building and Technology Assessment

I. Virgin, R. J. Frederick, and S. Ramachandran
Biotechnology Advisory Commission, Stockholm Environment Institute, Stockholm, Sweden

Background

The diversity of techniques that constitute modern biotechnology offers innovative solutions to today's pressing need for sustainable development in agricultural, environmental, and energy applications. Not surprisingly, many developing countries have shown great interest in establishing research and development programs in genetic engineering. Sharing the interest is, of course, not enough. The rate of development and the level of success are dependent not only upon the scientific and technical capabilities of resident scientists, but also on having a supportive infrastructure and an accepting environment in which to introduce and use biotechnologies. A key component in the formulation of a "biotechnology-accepting" environment is the establishment of a biosafety[1] regulatory oversight infrastructure (Brenner, 1995; Persley et al., 1992; Virgin and Frederick, 1995).

While biosafety issues have been discussed for over 20 years in the developed world, only in the last four years have the issues been extensively explored in the global arena. Apart from farmers, industry, and nongovenmental organizations (NGO)s, the Convention on Biological Diversity[2] and Agenda 21[3] program extensions also will apply pressure on developing countries to adopt the technology. Generally, the lack of regulations may lead to reluctance on the part of northern companies or agencies to use the new techniques in collaboration with partners in developing countries. In addition, within developing countries national scientists are reluctant to develop transgenic organisms without government sanction or regulatory reviews.

To facilitate the safe transfer of technologies and fully benefit from genetic resources and the rapid advances in modern biotechnology, developing countries need to formulate biosafety regulations. Concomitantly, and of equal importance, the countries must acquire the capacity to implement these regulations via scientifically based, environmental impact assessments. To achieve this, several things are needed, including training at all levels to address shortage of human resources, access to information, and international expertise.

The following is an attempt to paint, with a broad brush, the scene of the current biosafety capacity building activities. In addition, we also discuss the development of criteria on how to measure the impact and the effectiveness of given assistance. To present all the ongoing activities is neither necessary nor practical, so only representative organizations and activities are used to illustrate the discussion.

Developing Countries' Needs

Before discussing capacity building, it is useful to try to define what constitutes achieving the capacity to successfully inculcate biosafety programs into regulatory structures. The United Nations Environmental Programme (UNEP) definition of capacity building is "...the strengthening and/or development of human resources and institutional capacities. It involves the transfer of know-how, the development of appropriate facilities, and training in sciences related to safety in biotechnology and in the use of risk-assessment and risk-management" (UNEP, 1996). Already formulated constraints and obstacles also are acceptable points of departure for the discussion.

While the large majority of developing countries currently lack specific biosafety regulations, they do have laws regulating plant quarantine, animal health, and environmental protection (Virgin and Frederick, 1995). Many of these acts can be adapted to cover aspects of safety in biotechnology or be used as frameworks for future specific regulations. The more difficult situations will be those countries without functioning organization structures and human capacity to ensure adequate policy implementation. Below follows a more specific list of the most important problems as expressed by the participants at the African Regional Conference for International Co-operation on Safety in Biotechnology held in Harare in 1993.[4] We believe the following list also is applicable to other regions in the world:

- Insufficient capacity for enforcement of guidelines and/or regulations
- Need for training at all levels to address shortage of human resources
- Lack of formulated biotechnology and biosafety policies
- Formulation and implementation of guidelines and /or regulations
- Need for information collection and exchange (e.g., access to databases and knowledge of global development)
- Need for risk assessment research focusing on specific African circumstances
- Need for more facilities and equipment to carry out proper monitoring and risk assessment research
- Establishment of biosafety committees at the institutional, national, and regional levels
- Planning and adaptation of methods to monitor effects of field tests and ensure compliance with regulations
- Need for funding of safety issues as integral part of research and development projects
- Need for national and regional collaboration

International Activities in Biosafety Capacity Building

Even though much of the incentive to deal with these shortcomings has to come from the developing countries themselves, there is a clear need for continuing assistance from international organizations. With a majority of the world's countries still without regulatory mechanisms and implementing experience (Virgin and Frederick, 1995), there will be a continuing need for these activities in the foreseeable future. To support capacity building in biosafety, one can distinguish the following two areas of international activities, both of which should be developed, but not necessarily simultaneously—

- Short-term training in risk assessment and risk management
- Sustainable strengthening of infrastructure and expertise in biotechnology

Table 1. Some of the international organizations involved in biosafety activities

Intergovernmental:
- Organization for Economic Co-operation and Development (OECD)
- Commission of the European Communities (EC)
- Consultative Group on International Agricultural Research (CGIAR)
- Technical Center for Agricultural and Rural Co-operation (CTA)
- UN Organizations
  - United Nations Environment Program (UNEP)
  - United Nations Industrial Development Organization (UNIDO)
  - World Health Organization (WHO)
  - Food and Agriculture Organization (FAO)
  - Parties to the Convention on Biological Diversity

Independent/nonprofit:
- Biotechnology Advisory Commission (BAC)
- International Service for the Acquisition of Agri-biotech Applications (ISAAA)
- International Service for National Agricultural Research (ISNAR)
- International Academy of the Environment, Geneva (IAE)
- Agriculture Biotechnology for Sustainable Productivity (ABSP)

Donor organizations
- Netherlands Ministry of Foreign Affairs (DGIS)
- Royal Danish Ministry for Foreign Affairs/
  Department for International Development
- Deutsche Gesellschaft für Technische Zusammenarbeit (GTZ)
- Rockefeller Foundation
- Swedish International Development Cooperation Agency (SIDA)
- The World Bank
- United States Agency for International Development (USAID)

Environmental NGOs
- Genetic Resources Action International (GRAIN)
- Greenpeace International
- Rural Advancement Fund International (RAFI)
- Union of Concerned Scientists (UCS)
- Third World Network (TWN)
- The World Conservation Union (IUCN)
- World Wildlife Fund for Nature (WWF)

These groups provide advice on biosafety issues, support and/or organize training or scientific workshops, sponsor symposia, and produce a variety of published materials including reports, proceedings, newsletters and books. The list of organizations is not meant to be comprehensive, but rather to represent the array and variety of groups involved in biosafety.

The former line of work will be discussed in greater detail below. The latter line of work is derived from the fact that it is virtually impossible to regulate and assess the risk of a particular genetically modified organism (GMO) when competence in relevant fields of biotechnology is weak or lacking. A sufficient understanding of biotechnology is necessary. Without competence in the basic techniques, the development of a regulatory biosafety structure occurs in a vacuum. Therefore, the link between competence in biosafety and biotechnology should be recognized. A step-by-step approach, building competence by using and developing the techniques while interacting with regulatory authorities, will help to create confidence in decision. Any long-term assistance program of biotechnology capacity development must integrate capacity building in biosafety, a fact recognized by many donors active in strengthening human resources in these areas.

There are three important components of biosafety capacity building: training in biosafety implementation, risk assessment advice, and information services. There are today a substantial number of international organizations that are directly or indirectly providing assistance to developing countries within these fields of biosafety capacity building (Table 1). The extent of these efforts suggests closer co-ordination between providers of assistance and donors. However, the goal of a more harmonized and optimal provision of capacity building may have to await the formulation of a concrete action plan.

Training in Biosafety Implementation

Collaborative training programs for building capacity in the implementation of risk assessment and management have been sponsored by a variety of organizations. Table 2 contains a list of biosafety workshops that have taken place over the past three years. Designed to illustrate regulatory infrastructure and the means for implementing guidelines, many are offered at no cost to developing country scientists. A prime objective of these workshops has been to build institutional and individual capacity by sharing industrialized country experience in biosafety regulations and field testing of GMOs with scientists, policy makers, and special interest group representatives. In addition to workshops there have been discussions on internship/fellowship programmes. Trainees would spend an extended period of time (e.g., 4–8 weeks) within an experienced agency that reviews proposals or within a company that prepares applications and conducts field trials with transgenic organisms.

A typical biosafety implementation workshop as performed by the Biotechnology Advisory Commission (BAC) contains lectures on general principles and practices of biosafety (risk) assessment, risk management, overview of international development, monitoring, in-depth presentations by experienced regulators on practical implementation issues, and "hands-on experience" where the participants evaluate field trial proposals and are trained in making regulatory decisions. The latter type of exercises give trainees a better idea of what kind of information is needed for different types of regulatory decisions (i.e., limited field trials, commercialization of transgenic organisms). The overall goal is to give the participants enough knowledge and confidence so they can make a regulatory decision.

In other meetings policy makers and special interest groups have gathered to discuss common frameworks that can be developed to serve particular national needs within specific regions (ISAAA/USDA meeting, Table 2). Biosafety courses have been given by the United Nations Industrial Development Organization (UNIDO) and Agricultural Biotechnology for Sustainable Productivity (ABSP) on a yearly or semiannual basis.

Workshops on special topics (e.g., virus resistant plants and new plant viruses) have helped to define issues and refine supportable arguments critical to the determination of appropriate risk or concerns. Even when consensus could not be obtained, discussions often led to broader thinking of the issues and better appreciation for other points of view. The recent debates connected to the Convention on Biological Diversity have highlighted new issues. The uncertainty surrounding the environmental implications of growing genetically modified crops in their centers of origin is one important example. Recently, the Biotechnology Advisory Commission (BAC) and Inter-American Institute for Agricultural Advancement (IICA) organized a workshop to explore the issue in Latin America using transgenic potatoes as a model.[5]

Independent Risk Assessment Advice

Countries faced with their first application for environmental release of a GMO may experience difficulty in obtaining relevant information sources for their environmental impact assessment reviews. They also may have problems in interpreting this information in meaningful ways (i.e., in a fashion most useful for decision makers to choose between impartial options). Therefore, access to international expertise within the framework of an independent, impartial organization can be a valuable resource in understanding, adapting, and utilizing scientifically sound assessments. The BAC provides this type of service, and by including country representatives, the evaluation procedure provides capacity building through a participative process. There are also a number of bilateral support activities. One example is the information exchange between USDA/APHIS[6] and the Mexican National Committee for Agricultural Biosafety (CNBA). Another example is the recent bilateral agreement between United Kingdom and Argentina on safety in agricultural biotechnology.[7]

Biosafety Information

Information is being generated at nearly overwhelming rates, and individuals are being forced to specialize to keep up with the information flood. Since the early 1980s the number of data base systems has proliferated into thousands. There is a limited value in 400 pages of downloaded data (printed or electronic media) on the construction and use of transgenic insect resistant Bt-maize for an organization or country with limited scientific capacity and experience with gene technology. Many developing countries need impartial advice to develop useful connections for their scientific and technical environmental assessment. While knowledge of the critical issues covered in another country's environmental evaluations is helpful, it must be offered in such a way that a full appreciation of the process is possible.

There is a substantive body of information on biosafety regulations and experience with GMO's in North America and Europe (Table 3). The data included in databases are national biosafety regulations/guidelines, names of national biotechnology authorities and experts in risk assessment, and data on the field release and commercialization of genetically modified organisms. Most of these databases are available through the Internet or through experienced authorities (USDA/APHIS) or relevant advisory organizations.

Recognizing the Need for Evaluation Criteria

Capacity building is a knowledge based endeavor. One can describe the assistance in biosafety capacity building as a technology transfer process. In technology transfer, knowing what works and why are essential for success. Likewise, knowing what does not work is as important.

Table 2. A representative list of workshops and regional conferences that included biosafety (April 1993—September 1996)

Meeting	Organizer(s)	Date
Biosafety Workshop (Bogor, Indonesia)	ISAAA	April 1993
Latin America and Caribbean Region Biosafety Workshop (Oracabessa, Jamaica)	ABSP/(B/C)CRSP	May 1993
Environmental Impacts of Aquaculture Using Aquatic Organisms Derived Through Modern Biotechnology (Trondheim, Norway)	OECD	June 1993
ASEAN Biosafety Meeting/Workshop (Bangkok, Thailand)	UNIDO/NCGEB	November 1993
African Regional Conference for International Cooperation on Safety in Biotechnology (Harare, Zimbabwe)	RCZ/DGIS/VROM	October 1993
AGERI/ABSP/UNIDO Regional Biosafety Workshop (Cairo, Egypt)	AGERI/ABSP/UNIDO	January 1994
Commercialization of Agricultural Products Derived Through Modern Biotechnology (Washington, DC)	OECD	June 1994
Harmonization of Biosafety in the Americas: Building Institutional Capacities (Cartegena, Colombia)	IICA	September 1994
Biotechnology Regulation: Towards the Establishment of Intergovernmental Cooperation in Central and Eastern Europe (Vienna, Austria)	UNIDO	September 1994
Turning Priorities into Feasible Programs: Regional Seminar on Planning, Priorities and Policies for Agricultural Biotechnology in Southeast Asia (Singapore)	IBS/ISNAR	September 1994
Biosafety Training Workshop (Ibadan, Nigeria)	BAC/IITA	January 1995
Biotechnology and Biosafety: Benefit and Risk Assessment in an African Perspective (Nairobi, Kenya)	ISAAA/USDA	February 1995
Biodiversity and Harmonization of Biosafety in Central America and the Dominican Republic (San Jose, Costa Rica)	IICA/GTZ	February 1995
Turning Priorities into Feasible Programs: Regional Seminar on Planning, Priorities and Policies for Agricultural Biotechnology (South Africa)	IBS/ISNAR	April 95
Environmental Concerns With Transgenic Plants in Latin America: Potato as a Model (Puerto Iguazu, Argentina)	BAC/IICA	June 1995
Regional Consultations on Biosafety Technical Guidelines (San Jose, Bangkok, Amman, Buenos Aires, Geneva, Cairo, and Keszthely)	UNEP	March 1995 to September 1995
Biosafety Training Workshop for African Regional Biosafety Focal Point (ARBFP) Representatives	BAC/ARBFP	June 1996

Despite inherent difficulties, most would agree that measurements of aid effectiveness should be applied. However, when consideration moves to the question of how one measures and what evaluation criteria for efficiency should be used, difficulties arise. How broad should the areas of consideration be? Whose criteria should be used? How is consideration of all perspectives assured in the evaluation process? And, ultimately, how do we ensure the best information to make the best decisions? As with many technology transfer processes, the results may only be visible over the longer term. Nevertheless, it is possible to make reasonable informed adjustments to assistance efforts based on past experience and constructive evaluations. At a time when discretionary resources are diminishing and programs have to be adjusted to reflect economic downturns, it is also the time to consider how to be more effective in the pursuit of international biosafety assistance. Intergovernmental and nongovernmental organizations may work independently, but they probably compromise their efficiency and effectiveness when they work in isolation. Redundancies, missed opportunities for collaboration, and the potential stagnation of ideas and concepts may go hand in hand with a reluctance to interact.

Table 3. Representative databases containing biosafety information

Database	Location
BioTrack	Paris, France
Biosafety Information Network and Advisory Service (BINAS)	Vienna, Austria
GIBIP database	Basel, Switzerland
ICGEBnet	Trieste, Italy
International Initiatives in Agricultural Biotechnology: A Directory of Expertise	The Hague, Netherlands
National Biological Impact Assessment Program	Blacksburg, Virginia, USA
USDA Animal Plant Health Inspection Service	Riverdale, Maryland, USA
Information Resource for the Release of Organism Into the Environment	Cambridge, UK

Faced with these questions, and wishing to improve its biosafety assistance program, the BAC organized "Capacity Building in Biosafety for Developing Countries: Evaluation Criteria Development Workshop" in Stockholm, Sweden

Table 4. Examples of criteria for succesful biosafety training

Questions	Indices of Positive Action
1. Did the efforts match the need?	1. a) Number of persons trained proportional to status of policy and regulatory developments, anticipated number and timing of requests for GMO release, and level of in-country research. b) Local input sought to identify subject matter and presentation appropriate to participants need (choice of topics, degree of specificity).
2. Were the right people chosen and were criteria for selection of trainees justified?	2. Trainees situated in relevant positions (e.g., in biosafety committees or regulatory authorities).
3. Was the training directly utilized?	3. a) Trainees participated in development of national regulatory structure. b) Trainees serve as national experts or in biosafety committees. c) Trainees disseminate information or act as resource persons in national or local training programs
4. Were the training workshops timely?	4. a) Trainees able to use their knowledge immediately after the workshop (and not several years later). b) Trained personnel available at the time applications are made. c) Sufficient capacity to handle application load. d) Training relevant for current stage of country development (e.g., formulation of implementation of regulations, monitoring of GMOs)
5. Were training information and resource materials up to date?	5. Trainees able to use state-of-the-art scientific and technical information derived from countries with more experience and higher level of activities.
6. Did the training improve biosafety skills and knowledge to the stated goal?	6. a) Trained individuals sufficiently competent and confident to make regulatory decisions. b) The country able to do evaluations independent of outside support.
7. Are the trained persons using or disseminating information or acting as functional resource persons in national training programs?	7. a) Follow up seminars by trainees at national and institutional level being held. b) National and institutional biosafety training workshops, policy related workshops being supported. c) International networks accessible.

in May 1996.[8] At the workshop the experience of organizations providing biosafety assistance and the recipients' needs could be shared. One of the workshop objectives was to investigate possibilities for co-operation in biosafety capacity building. The second workshop objective was to use the collective expertise and experience to develop acceptable and objective evaluation criteria to monitor the effect of given assistance programs. Carefully crafted, the evaluation criteria will stimulate discussion among the involved parties on how to improve the effectiveness of capacity building in biosafety. They will enable providers and recipients of assistance, as well as selected donor agencies to draw informed conclusions regarding the potential impact of initiatives and activities. Finally, they could be used for the incremental adjustments of assistance programs increasing their efficiency.

Experience With Evaluation of Developmental Assistance Programs

In a 1995 report, Moore, Stewart, and Huddock presented a summary of the literature and ideas surrounding institutional building as a development assistance method. The contents of the report are relevant to a broad range of development assistance programs. To the extent that biosafety capacity building is analogous to an institutional building process, this report provides a useful introduction for creating evaluation criteria lists. As these authors point out, general checklists of measures for success (e.g., growth of professionalism, development of linkages, organizational independence) differ widely. Moore and colleagues describe several areas for evaluation that can be utilized for biosafety capacity building:

- *Objectives:* the extent to which the organization has objectives for the institutional building process which are both clear and agreed upon with relevant other parties and the extent to which it is moving towards achieving them.
- *Resources:* the sources and adequacy of financial and personnel resources.
- *Political support:* the extent of political support that an institution enjoys, its consequences, to what extent such support is needed.
- *Internal structure and functioning:* internal organization, the ability to implement regulations, and human resources development.
- *Outputs and outcomes:* the extent to which the outputs or outcomes of the organization's activities can be measured, and success in achieving them.

The Development of Evaluation Criteria

The BAC workshop resulted in a detailed list of evaluation criteria. The full criteria table includes five different categories: development of national policies, development of biosafety regulations, implementation of biosafety programs, biosafety assessment and monitoring research, and capacity building in biotechnology. In this paper we have limited ourselves to list some of the criteria relevant to biosafety training. Table 4 is presented under a two column heading. The first column is a list of questions that might be asked in an evaluation process. Each question implies defining an event or events that are either necessary or recognized as an important part of biosafety capacity building. Associated with each type of event may be specific type of assistance, (e.g., training workshops, and information exchange). The second column is a list of corresponding indices of specific indicators that might be used in responding to the questions. In most cases, the signs of a successful capacity building program would be found in how complete or comprehensive the answers to the questions may be. Admittedly, this approach can be criticized as being subjective, especially if it is viewed merely as a checklist. For example, having a workshop may be considered a positive response, but if it is poorly done, its value to the overall building process may be nil. Consequently, the criteria should be used with some caution, preferably within the context of a national or regional plan. It can not be used as an alternative to evaluation of independent events on their own merits.

Concluding Remarks

There are many activities ongoing that can be used to further biosafety capacity building and it is incumbent upon country representatives to take advantage of them. To be most helpful, perspectives should be broad enough to include not only biosafety evaluations, but also monitoring; information collection, storage and exchange; and the rights of parties (e.g., patent holders, farmers, indigenous peoples). The challenge of inculcating biosafety into a global society remains a daunting project. It will require participation by all stake holders—researchers, developers, government authorities, and the public. Progress and process must be critically reviewed and evaluated to allow for changes to accommodate the rapid evolution of biotechnology. With continuing effort from international organizations, biosafety will evolve with the technology as it spreads throughout the world.✧

Footnotes

1. *Biosafety is here defined as the policies and procedures adopted to ensure the environmentally safe applications of biotechnology.*
2. *As a result of the concerns regarding the safe use of biotechnology, the Second Conference of Parties (COP2) in Jakarta in 1995 decided to develop a biosafety protocol. (UNEP/CBD/COP/2/CW/L.22).*
3. *As a follow up to Agenda 21, UNEP has hosted several regional consultations of government—designated experts and developed Voluntary International Technical Guidelines for Safety in Biotechnology. The next phase is the development of UNEP/UNDP/World Bank mediated biosafety capacity building programs funded by the Global Environmental Facility (GEF).*
4. *Organized by the Research Council of Zimbabwe and the Netherlands Ministries of Foreign Affairs and of Housing, Physical Planning and Environment.*
5. *Frederick, Robert J, Virgin, Ivar and Lindarte, Eduardo (eds.) (1995) Environmental Concerns with Transgenic Plants in Latin America: Potato as a Model. Proceedings of a BAC/IICA workshop, 90 pp. Publisher: Biotechnology Advisory Commission.*
6. *United States Department of Agriculture/Animal and Plant Health Inspection Service.*
7. *The competent authorities in the two countries (CONABIA and Department of Environment, UK) exchange expertise and biosafety information.*
8. *The proceedings of the conference containing a detailed evaluation criteria list is published and could be ordered from the Biotechnology Advisory Commission (BAC).*

Selected Reading

Brenner, C., 1995. Technology transfer: public and private sector roles. In: J. Komen, J. Cohen, and S.K. Lee (eds.), Turning priorities into feasible programs: proceedings of a regional seminar on planning, priorities and policies for agricultural biotechnology in Southeast Asia, 134 pp. The Hague/Singapore: Intermediary Biotechnology Service/Nanyang Technological University.

Frederick, Robert J., 1995. International activities in support of safety in biotechnology. In: Proceedings of a Central and Eastern European conference for regional and international cooperation on safety in biotechnology, p. 65–74. September 4–6, 1995, Keszthely, Hungary.

Frederick, Robert J., 1995. Biosafety information capacity building, p. 7–8. In: Biosafety Network Newsletter, Regional Biosafety Focal Point, No. 2. Harare, Zimbabwe.

Persley, G.J., Giddings, L.V., and Juma, C., 1992. Biosafety: the safe application of biotechnology in agriculture and the environment. The Hague: International Service for National Agricultural Research.

Virgin, I. and Frederick, R.J., 1995. The need for international harmonization of biosafety regulations. African Crop Science Journal 3:(3)387–395.

Virgin, I. and Frederick, R.J. (eds.), 1996. Biosafety capacity building: evaluation criteria development. In: Proceedings from a BAC workshop. Biotechnology Advisory Commission, ISBN: 9–188–71429–2, (in press).

Genetically Modified Organisms—Biosafety Implications for India

Ghayur Alam
Centre For Technology Studies, Hauz Khas, New Delhi

Introduction

Genetically modified organisms (GMOs) were first developed and released for trial in the early 1980s. Since then, a large number of GMOs have been released for field trial in the United States (U.S.), Japan, China, and countries belonging to the European Community (EC). More than 2,500 cases of experimental release of genetically engineered organisms were known in the U.S. by May 1, 1995.[1] Fifteen transgenic plants and 4 altered microorganisms have already received marketing permission. Another 13 transgenic plants and two microorganisms were waiting for permission in November 1996.[2]

The EC countries that have been comparatively less active were reported to have released 496 transgenic plants in 1986–1994. Of these, France, the Netherlands, Belgium, and the United Kingdom (UK) were most active (168, 84, 81 and 78 releases, respectively).[3] By 1994, the EC had given marketing permission for four GMOs: three vaccines (pseudorabies and fox rabies) and one herbicide tobacco.[4] In Japan transgenics of 20 crop plants were reported to have been field tried between 1988 and January 1995.[5] China is known to be undertaking the development of GMOs on a very large scale. According to reports published in *Science*, transgenic plants (mostly with virus resistance) are being planted on hundred of thousands of acres. Most of these trials involve tobacco plants. In fact, cigarettes with transgenic tobacco are already being marketed in China.[6] Other countries with a significant number of field trials involving GMOs are Canada (359 between 1986–1994), Argentina (43), Mexico (20), Chile (17), Australia (26 between 1986–1994) and New Zealand (15 between 1988–1994).[7]

GMOs and Biosafety

GMOs have a number of biosafety implications. Some of the important potential risks are

1. There is a danger that the new genes incorporated in these plants may enhance their ability to survive in adverse conditions to such an extent that they become weeds. Some of the properties which could contribute to weediness in a plant are tolerance of adverse environmental conditions (e.g., salt, temperature, and moisture) and resistance to pests such as insects and viruses.[8]

   The large-scale release of transgenic plants also involves the possibility of the transfer of the transgenes to other plants. As a result, the latter can become weeds. According to some experts, the possibility of gene flow and its consequences are the biggest concerns raised by field trials of genetically modified plants.[9]

2. GMOs with insecticidal properties such as baculoviruses may attack nontargetted species of pests. Take, for example, a genetically improved baculovirus (NPV) of the alfa-alfa looper which contains an insect specific neurotoxin from the venom of a North African scorpion, developed at the Institute of Virology and Environmental Methodology (University of Sussex). (American Cynamide also has developed similar baculoviruses). A proposal to carry out field trials of the virus evoked strong protest from environmentalists, who feared that the virus is not specific enough and may kill a number of nontargeted insects, including some rare moths.[10]

3. The widespread use of plants with insecticidal genes may cause (or accelerate) the development of resistance to the insecticide involved. The development of transgenic plants with insecticidal genes is one of the most common objectives of research in this field. For example, about 22% of the transgenic crops involved in field trials in the U.S. between 1987 and 1995 were transformed for insect resistance.[11] While insect resistant cotton, corn and potato are already approved for marketing in the United States (U.S.), a number of other insect-resistant crops are in the pipeline. The planting of these crops on a large scale may aggravate the problem of resistance to insecticides.

Most transgenic plants with insect resistance have genes incorporated in them to produce what is generally called Bt toxin.[12] For example, of 349 applications and notifications submitted to the U.S. Department of Agriculture between 1987 and 1995 to field test transgenic plants with insecticidal genes, 244 (70%) contained Bt genes.[13] Bt corn (Ciba-Geigy; Mycogen), cotton (Monsanto) and potato (Monsanto) have already received marketing approval in the U.S., and large-scale planting of Bt cotton is already taking place. For example, Monsanto's Bt cotton has been planted on 2 million acres in the US.. In fact, about 13% of the U.S. cotton planting this year was of a transgenic type with Bt genes.[14]

The implications of large-scale use of Bt crops for the development of resistance have been a focus of intense debate among biotechnologists and environmentalists. Most companies with Bt crops use a resistance management strategy which consists of a) the production of high doses of Bt toxin and b) planting a part of the field with non-Bt plants (called refugia).[15] It is expected that the high dose of Bt toxin will kill most of the targeted insects. Furthermore, any surviving insects will be likely to mate with insects from the refugia (insects which are not exposed to Bt toxin). This will prevent the development of insects with a high degree of resistance to Bt toxin.

The environmentalists feel that the resistance management strategies being adopted for Bt crops are not satisfactory. They are particularly concerned that the Bt plants may not produce a sufficiently large quantity of Bt toxin to kill most insects. The companies, on the other hand, claim that their Bt plants produce a high enough dose of Bt toxin to kill all the pest insects. Mycogen, for example, claims that its transgenic corn produces Bt at levels 20 times higher than that needed to kill the European corn borer. It is hoped that this will kill all the pests and none will survive to develop resistance.[16]

However, according to the critics, the plans submitted by the companies provide little evidence from field experiments to confirm that the Bt doses produced are high enough to be suitable for a high dose plus refugia strategy.[17] They further argue that even if a small section of the insect population survives Bt toxin, resistance will develop rapidly.[18] Also, there is a possibility that the production of Bt may not be uniform. If some plants fail to produce sufficient Bt, resistance will develop fast.

Recent reports suggest that these fears are not unfounded. Bt cotton plants, when planted on a large scale, have not behaved as expected. Up to 20,000 acres (out of 2 million acres) of Monsanto's Bt cotton is reported to be under attack by cotton bollworm.[19] Although the precise cause for this is in not known, it is suspected that the plants are not producing enough Bt toxin to kill most of the pests. In that case, there would be a distinct possibility that some insect pests would survive and would build resistance.

4. Large-scale use of transgenic virus resistant plants may lead to the development of new viruses. The first virus resistant transgenic crop was given marketing permission in 1995 in the U.S. after strong protests from the environmentalists.[20]

Biosafety Regulations in Developed Countries

The steps to regulate the release of GMOs were first taken in developed countries in the 1980s. For example, a biotechnology risk assessment research program was launched by the U.S. Environmental Protection Agency (EPA) in 1986.[21] The OECD also began to focus on the safety aspects of biotechnology in the mid-1980s. Since then most developed countries have formulated rules to regulate the use and release of GMOs.

Recently, there has been a trend towards relaxation of biosafety rules in developed countries. A number of them [such as the United Kingdom (UK), Germany, and the United States] are in the process of relaxing the rules regulating the release of GMOs.[22] USDA, for example, decided in August 1995 that most field tests of genetically engineered crops in the U.S. would be exempted from risk assessment.[23] The UK and Germany have already relaxed regulations to make field trials easier.[24] The EC is planning to relax the regulations, which require companies and research institutions to carry out an impact assessment and get approval from governments before releasing GMOs.[25]

The relaxation of biosafety regulations in these countries is partly a reflection of the increased confidence of regulatory authorities in the safety of GMOs. It is, however, also dictated by the fear that strict regulations will slow down the development of local biotechnology industries and make them uncompetitive.[26]

International Biosafety Protocol

A legally binding international biosafety protocol is desirable for a number of reasons. The most important of these are: a) it will ensure that countries do not neglect biosafety requirements in order to promote their biotechnology industries, b) it will prevent the testing and use of GMOs in developing countries with weak or no regulations,[27] and c) it will help the developing countries to formulate national biosafety regulations. They can learn from the experience of developed countries.

The need for international biosafety guidelines was accepted at the 1992 Rio Earth Summit. After initial opposition from some developed countries (especially the United States, Germany, Japan and Australia) the need for an international protocol is widely accepted.[28] There are, however, serious differences about the scope of the protocol. Developing countries, by and large, prefer a comprehensive protocol, covering research, transport, and use of the GMOs. Many developed countries, on the other hand, prefer a protocol with its scope limited to the transport of GMOs.[29] The developed countries also oppose a strict protocol as they fear that this will affect the development of biotechnologies in their countries.[30]

A compromise between the G77 and developed countries was reached in 1995. According to this the protocol will regulate the transport of GMOs and not their use.[31] However, a meeting of the ad hoc group of experts on biosafety to prepare a draft of the protocol, held in July 1996, failed due to serious differences between developed and some developing countries. While a group of developing countries (Malaysia, Ethiopia, and Mauritius) wanted to expand the coverage of the protocol to include the use of GMOs, the developed countries and the biotechnology industry opposed this.[32]

A draft text of the biosafety protocol is expected to be ready by 1998. However, given the serious differences of opinion between various groups of countries, its preparation may be delayed. Moreover, the final protocol may be a weak document, limited to the transboundary movement of GMOs.

In the absence of a comprehensive international protocol, the use of GMOs will need to be regulated by national biosafety regulations.[33] Presently, very few developing countries have such regulations. These include China, India, and the Philippines.[34] Some Latin American countries are in the process of preparing biosafety regulations.[35]

Biosafety Regulations in India

Biosafety regulations were first prepared in India in 1990. These were revised in 1994. The regulations cover both the research and field trials of GMOs. Furthermore, they cover both the locally developed GMOs and imported GMOs.

The regulations are implemented by three committees. These are as follows:

1. Institutional Biosafety Committee (IBSC). All institutions intending to undertake research activities involving genetic manipulation of microorganisms, plants, and animals are required to set up IBSCs. IBSC includes representatives from the concerned institution, the Department of Biotechnology (DBT) and a doctor. At present 69 research institutes have IBSCs. The researchers are required to inform the institute's IBSC of any experiment with biosafety implications. In addition to looking after the biosafety aspects of experiments, the IBSC is also responsible for the training of researchers on biosafety and for monitoring the health of researchers.

2. Review Committee on Genetic Manipulation (RCGM). The RCGM functions under the DBT and monitors all ongoing research with biosafety implications. The committee also issues clearance for import and export genetic material required for research and training purposes.

3. Genetic Engineering Approval Committee (GEAC). GEAC functions under the Ministry of Environment and is responsible for biosafety in the cases of:

   a. Field trials and large-scale (commercial) release of GMOs

   b. Production, sale, import, or use of genetically engineered products

   c. Import, export, transport, handling, and use of GMOs

In addition to these, there are also committees at the state and district level.

According to the perceived risks involved, the genetic experiments are classified into three catagories. The experiments considered to have the highest risk include the use of toxin genes, cloning of genes for vaccine production, transfer of antibiotic resistant genes to pathogenic organisms not normally carrying such resistance, manipulation involving plant and animal viruses, and gene transfer to whole plants and animals. These experiments require clearance by the highest expert committee on a case-by-case basis.[36]

A complete list of research with biosafety implications being carried out in India is not available publicly. It is, therefore, not possible to make definite comments on the potential risks associated with research on GMOs in India. Our information, however, suggests that most of the research on GMOs in India is at an early stage.[37] This includes the following:

1. Tagging and transfer of blight resistant genes from wild to cultivated species of rice
2. Cloning of genes responsible for resistance to gall midges in rice
3. Development of transgenic rice with Bt genes
4. Development of transgenic rice with fungicide tolerance
5. Development of transgenic rice with herbicide resistance
6. Development of transgenic chickpea with Bt genes
7. Development of transgenic cotton with Bt genes.

As most of the above mentioned research is at a very early stage, field trials are unlikely to take place in the near future. The only field trials currently taking place in India involve imported transgenic plants. The Annual Report of the Department of Biotechnology (1995–96) has listed two of these. (We are not aware of any others.)

One of these involves small-scale field trials (on a 5 meters by 5 meters plot) of transgenic rapeseed canola (hybrid) and Bt tomato by Proagro. In both cases the transgenic material has been imported from Plant Genetic Systems (PGS International) of Belgium.[38]

In the other case, MAHYCO, a large Indian seed company, has proposed to undertake trials in glass houses of imported Bt cotton by crossing it with Indian lines. Currently, MAHYCO is field testing Bt cotton in more than 40 locations in India. Similarly, Proagro Seed Company has also field tested genetically engineered canola.

While locally developed GMOs are unlikely to be ready for field trial, more imported transgenics may enter India in the near future. India's seed policy, which was liberalized in 1988, allows free import of seeds and other planting materials. Moreover, with the relaxation in the rules governing foreign collaborations and investment, a number of seed multinationals such as Cargill, Pioneer, Northrup, Asgrow, Continental Grains and PGS are already active in India. Once India enacts some form of plant variety protection legislation, more seed and agrochemical multinationals can be expected to enter the Indian market.[39] It is likely that some of these will introduce transgenic crops.

Conclusions

India is better prepared than most developing countries to deal with the problem of biosafety. It has a well defined system and a set of guidelines to monitor and regulate the release of GMOs. However, the system is yet to be tested in practice.

India's experience with GMOs is almost exclusively confined to laboratories. We have no experience of dealing with the release of GMOs on a large scale. The experience with Monsanto's Bt cotton in the U.S. shows that the behaviour of transgenic plants, when planted on large areas, is far more complex and unpredictable than when they are tried on a small scale. Lacking experience with large-scale release of GMOs, India can face serious biosafety problems if and when large scale trials and commercial releases take place. A large body of knowledge and experience already exists in developed countries on regulating large-scale field trials and commercial releases of GMOs. It is important that the Indian system be capable of absorbing this knowledge and experience. However, even the regulations in force in developed countries may not be adequate. As mentioned above, regulations have been relaxed in many countries in order to promote local biotechnology industries. Therefore, while learning from the experience of developed countries, India will need to maintain its own regulations. Moreover, it is important that the system be flexible enough to absorb new knowledge.

We would like to conclude by pointing out two serious weaknesses in India's biosafety regulatory system.

First, there is a lack of public participation in the process. The information regarding research and field trials with possible environmental and other risks is not made public. In the absence of detailed information it is impossible to judge the effectiveness of India's biosafety regulations and their implementation. In order to increase the credibility and accountability of the system, it is necessary that more information be made available to the public.

Second, the implementation of biosafety guidelines are likely to pose serious problems. This will be particularly difficult when transgenics are planted by farmers. In many instances, the safe use of these crops will require that certain conditions (such as leaving a part of the plant as refugia in the case of Bt crops) are followed. The past experience with pesticides shows that the guidelines and instructions suggested by the government and manufacturers are often ignored by farmers. If this happens in the case of transgenic plants, the consequence could be serious. It is, therefore, clear that the formulation of adequate biosafety regulations is not enough. We must also ensure that these regulations will be followed at the farm level.✧

Footnotes

1. *Gene Exchange, Vol. 5, No. 4; Vol. 6, No.1, July 1995, p. 16.*
2. *What is coming to market? An Update on commercialization, Gene Exchange, Vol. 7, No. 1, December 1996, p. 4–5.*
3. *Gene Exchange, December 1994, p. 7.*
4. *Bullard, 1995, p. 12.*
5. *Commandeur, 1995, p. 9–12.*
6. *Plafker, 1994, p. 966.*
7. *Gene Exchange, December 1994, p. 7 and Kathen, 1996, p. 13.*
8. *Rissler and Mellon, 1993, p. 17–24.*
9. *Ahlgoy and Duesing, 1996, p. 39–40.*
10. *Agbiotech News and Information, September 1994 and May 1995.*
11. *Gene Exchange, July 1995, p. 11.*
12. *The toxin, which has insecticidal properties, is produced by soil bacterium, Bacillus thuringiensis, and is widely used (as a spray) by organic farmers as a biocontrol agent. In fact, it is the largest selling biocontrol agent, with annual global sales of more than U.S. $125 million. See: Tabashink, 1995, p. 24.*
13. *Insects adapt to protease inhibitors in transgenic plants. The Gene Exchange, Vol. 6, No. 2 and 3, December 1995, p. 6.*
14. *Kaiser, 1996, p. 423.*

15. *Managing resistance to Bt. The Gene Exchange, Vol. 6, No. 2 and 3, December 1995, p. 4–7.*
16. *Fox, 1995, p. 1035–36.*
17. *Managing resistance to Bt. The Gene Exchange, Vol. 6, No. 2 and 3, December 1995, p 5. Bruce Tabashink of the University of Hawaii, who was among the first researchers to detect Bt resistance in the field, is skeptical of the Bt resistance management strategy. According to him, the strategy is based on a lot of theory and not enough experience. See Hoyle 1996, p. 803.*
18. *Resistance to Bt toxin is already known to exist in the diamondback moth in Hawaii, USA, Japan, Malaysia, Philippines, Thailand, and China. Laboratory research has shown that the potential of Bt resistance is widespread among insects. See Tabashink, 1995, p. 24–25. Furthermore, the U.S. Environmental Protection Agency (EPA) is reported to be of the view that if Bt plans are used in large numbers, resistance may develop within 3–5 years. See "What is the rush?" (editorial). Gene Exchange, Vol. 6, No. 2 and 3, December 1995, p. 2.*
19. *Kaiser, 1996, p. 423 and Maciwain, 1996, p. 289.*
20. *Gene Exchange, August 1994, p. 9.*
21. *Gustaffson, 1994, p. 236.*
22. *Hodgson, 1995, p. 1035.*
23. *Fox, 1995, p. 1035. USDA proposes to abandon oversight of field testing. Gene Exchange, Vol.6, No. 2 and 3, December 1995, p. 3. Similar relaxation of biosafety regulations has been made in the case of medical biotechnology. For example, USFDA has decided to treat biotech drugs the same as chemically synthesized drugs, needing no special approval. See Kleiner, 1995, p. 12.*
24. *Abbot, 1994, p. 94.*
25. *Mackenzie, 1994, p. 6.*
26. *Coghlan and Science, Vol. 267, 20 January 1995, p. 326.*
27. *In a number of cases, field testing of GMOs is reported to have been performed in countries with weaker regulations by researchers from Europe and the USA. See Science and Public Policy, Vol. 16, No. 4, August 1989, p. 211–233.*
28. *According to the U.S. Biotechnology Industry Organization, "the inevitability of a protocol is now accepted by everyone." See Hoyle, 1996, p. 803.*
29. *Masood, 1996, p. 384.*
30. *Dickson, 1995, p. 94.*
31. *Masood, 1995, p. 326.*
32. *Masood, 1996, p. 384.*

33. *Some developing countries have signed bilateral biosafety agreements with the UK and Netherlands. This is expected to help these countries in receiving MNC investments and technology. See Dickson, 1995, p. 94.*
34. *Although China introduced biosafety regulations in 1993, neglect of biosafety is reported to be common. See Plafker, 1994, p. 966–967 and Bijman, 1994.*
35. *Commandeur, 1996, p. 22.*
36. *India 1994, p. 3.*
37. *Alam, 1994, p. 30–36.*
38. *India, 1996, p.110. Proagro Seed Company of India has a joint venture with Plant Genetic Systems (PGS International) of Belgium to develop, produce, and market hybrid oilseed rape (canola) and hybrid vegetable seeds for the Asian market. The joint venture is known as Proagro-PGS India (pvt.). The production will be based on PGS's proprietary SeedLink hybridization technology. See Agbiotech News and Information, March 1994.*
39. *According to recent reports, the Indian government is proposing to introduce a Plant Varieties Protection Bill in the winter session of the parliament. See Economic Times, November 8, 1996.*

Selected Reading

Abbot, Alison, 14 March 1996. Transgenic trials under pressure in Germany, p. 94. Nature.

Ahlgoy, Patricia and Duesing, John H., January 1996. Assessing the environmental impact of gene transfer to wild relatives. Bio/technology 14:39–40.

Alam, Ghayur, 1994. Biotechnology and sustainable agriculture: lessons from India, p. 30–36. Technical papers, No. 103, OECD Development Centre.

Bijman, Jos, March 1994. Biosafety regulations. Biotechnology Development and Monitor, No. 18.

Bullard, Linda, December 1995. Update on deliberate release in the European Union, p. 12. Gene Exchange, Vol. 6, No. 2 and 3.

Commandeur, Peter, March 1995. Public acceptance and regulation of biotechnology in Japan. Monitor 22:9–12.

Commandeur, Peter, March 1996. North-South America Conference on Biotechnology. Biotechnology and Development Monitor 26:20–22.

Dickson, David, September 1995. Biosafety code gathers pace through bilateral agreements. Nature 377:94.

Fox, Jeffre L., October 1995. EPA okays Bt corn; USDA eases plant testing. Bio/technology 13:1035–1036.

Gustafsson, Kersti and Jansson, Janet K., June 1994. Ecological risk assessment of the deliberate release of gentically modified microorganisms. Ambio, Vol. 22, No. 4.

Hodgson, John, October 1995. Field testing following notification. Bio/technology, Vol. 13.

Hoyle, Russ, July 1996. Biosafety protocol draft spooks US biotechnology officials. Nature Biotechnology 14:803.

India, Government of, May 1994. Revised guidelines for safety in biotechnology, p. 3., Department of Biotechnology.

India, Government of. Annual Report 1995–96. Department of Biotechnology, 1996. p. 110–111.

Kaiser, Jocelyn, July 1996. Pests overwhelm Bt cotton crops. Science 273:423.

Kathen, Andre de, September 1996. The impact of transgenic crop releases on biodiversity in developing countries. Biotechnology and Development Monitor 28:13.

Kleiner, Kurt, November 1995. No more special safeguards for biotech drugs, p. 12. New Scientist.

Maciiwain, Collin, 25 July 1996. Bollworms chew hole in gene-engineered cotton. Nature 382:289.

Mackenzie, Debora, 11 June 1994. Europe pushes for a genetically modified future, p.6. New Scientist.

Masood, Ehsan, 23 November 1995. Biosafety rules will regulate international GMO transfers. Nature 378:326.

Masood, Ehsan, 1 August 1996. Liability clause blocks talks on biosafety protocol. Nature 382:384.

Plafker, Ted, 11 November 1994. Genetic engineering: first biotech safety rules don't deter Chinese efforts. Science 266:966–967.

Rissler, Jane, and Mellon, Margaret, December 1993. Peril amidst the promise— ecological risks of transgenic crops in a global market. Union of Concerned Scientists.

Tabashink, Bruce E., August 1995. Resistance to insecticide bacillus thuringiensis, and transgenic plants, p. 24–25. Pesticide Outlook, Cambridge.

Environmental Impact of Crops Transformed With Genes from Bacillus thuringiensis (Bt) for Insect Resistance

Michael B. Cohen
Entomology and Plant Pathology Division, International Rice Research Institute, Philippines

Abstract

Three species of insect-resistant transgenic crops with δ-endotoxin genes from *Bacillus thuringiensis* (Bt) were released to farmers in the USA in 1996. These and several other Bt crops will likely be released to farmers in many additional countries over the next few years. Insecticide products containing Bt have been widely used in agriculture for more than 30 years and have an excellent safety record. Researchable questions remain about the environmental impact of Bt crops, differing for particular crop species and geographic areas. These questions include indirect effects on the management of nontarget pest species by naturally occurring biological control, the fate and persistence of δ-endotoxins in the soil, the consequences of outcrossing of δ-endotoxin genes to wild and weedy crop relatives, and the evolution of pest resistance to δ-endotoxins. Risks associated with the use of Bt crops should be balanced against potential benefits, notably a reduction in the use of chemical insecticides in some crop systems.

Introduction

Bacillus thuringiensis (Bt) is a bacterium abundant in soils, grain dust, and other habitats in most parts of the world (Meadows, 1993). Upon sporulation, it forms a parasporal crystal composed largely of proteins known as δ-endotoxins, which are toxic to some orders of insects. Bt spore/crystal mixtures are in wide use as "microbial insecticides" for the control of agricultural and forest pests and vectors of human and animal pathogens and have an excellent safety record. Because of their safety and effectiveness, δ endotoxins were among the first proteins used for production of transgenic crop plants with resistance to insects.

After an extended period of research, development, and regulatory scrutiny, commercial production of Bt crops (maize, potatoes, cotton) began in the United States in the 1996 growing season. Fourteen additional crop species have been transformed with Bt genes, of which eight have been field tested (Stewart et al., 1996). In this article I review issues of environmental safety associated with field testing and commercial release of Bt crops. Other reviews of this topic include Jepson et al. (1994), describing microbial and invertebrate test systems useful for evaluating the environmental impact of Bt crops, and Hokkanen and Wearing (1994), summarizing the conclusions of a workshop on Bt crop deployment.

Diversity of Bt Strains and Toxins

A large diversity of *B. thuringiensis* strains, toxins, and toxin combinations produced by these strains, has been documented (Schnepf, 1995). Several dozen Bt subspecies have been described and are distinguished on the basis of serotyping and biochemical properties. A given Bt strain may produce one or several δ-endotoxins, or none at all. More than 50 members of the δ-endotoxin gene family have been cloned and sequenced. The genes and proteins are named with the prefix *cry* and Cry (from the word "crystal"), respectively, and are grouped into subfamilies on the basis of similarity in amino acid sequence. The members of each subfamily generally share toxicity to one or more insect orders. Genes of the *cryI* and *cryIII* subfamilies have been used in transgenic crops. The proteins encoded by these genes are toxic to larvae of some species in the order Lepidoptera (moths and butterflies) and adults and larvae of some species in the order Coleoptera (beetles), respectively.

Bt strains may produce several other toxins in addition to δ-endotoxins. Notable among these are β-exotoxins, nucleotide analogs that are broadly toxic to vertebrates and invertebrates. In many countries, strains producing β-exotoxin are prohibited from use in commercial Bt spray products (Meadows, 1993).

Impact of Bt Crops on Wildlife and Beneficial Arthropods

Because of the long and widespread use of Bt spore-crystal mixtures as topically applied insecticides, there is an extensive literature on the health (Siegal and

Shadduck, 1989) and environmental (Meadows, 1993) safety of Bt spray formulations. These materials are toxic to vertebrates only under extreme conditions, such as the injection of very large doses into lab animals (Siegal and Shadduck, 1989). The selective toxicity of Bt insecticides also makes them generally compatible with biological control of insect pests by natural enemies such as spiders and insect predators and "parasitoids" (parasitic wasps and flies that eventually consume and kill their hosts). Tests of Bt spore/crystal mixtures have shown only occasional direct toxicity to parasitoids, predators, and honeybees (Flexner et al., 1986).

δ-Endotoxins produced by Bt cells are aggregated in the form of crystals and must first be solubilized in the insect gut, then (in the case of CryI toxins) cleaved by proteases to yield an active fragment. By contrast, toxins produced in plants are already solubilized and, because only the DNA coding for the active fragment is used in plant transformation, already activated.

Because the form of δ-endotoxins produced in transgenic plants differs slightly from that of the native toxins, additional evaluation of the health and environmental effects of these modified proteins has been necessary. Documents compiled by seed producers of transgenic plants and filed with agencies responsible for the regulation of transgenic plants in the United States (U.S. Department of Agriculture, Animal and Plant Health Inspection Service [USDA-APHIS]; the Environmental Protection Agency; and the Food and Drug Administration [FDA]) have included tests of oral toxicity to rodents, birds, honeybees, insect predators and parasitoids, earthworms, and *Daphnia* (e.g., Ciba-Geigy, 1994; Monsanto, 1994). The FDA has ruled that products derived from Bt cotton, potato and maize can be used for human and animal consumption in the U.S. (S. Shantharam, USDA-APHIS, personal communication).

The cultivation of some Bt crops should lead to a substantial reduction in applications of synthetic insecticides currently directed against pest species of Lepidoptera or Coleoptera. This should reduce environmental damage caused by insecticide drift and runoff and also should enhance the naturally occurring biological control of pests that are not controlled by Bt crops, such as mites, aphids, and thrips. These taxa are often "secondary pests" that become problems as a result of the destruction of their natural enemies by insecticides directed against primary pests. In field trials in the U.S. and Canada, populations of several predator species were higher in plots of Bt potatoes (resistant to Colorado potato beetle) than in plots of conventional potatoes treated with insecticide for beetle control (Monsanto, 1994). In the Bt potato plots, insecticide applications were also no longer necessary against potato aphids, presumably because of enhanced biological control of these insects.

The effect of Bt plants on biological control of insect pests may be different in agroecosystems where insecticide use is less intensive than in cotton or potatoes and where biological control is already highly effective (such as rice in tropical Asia). Bt rice is being developed primarily for control of stem boring caterpillars (Wünn et al., 1996; IRRI, 1996). In many regions, stem borers are chronic pests that cause substantial area-wide, cumulative yield loss rather than large losses in the fields of individual farmers. Stem borers are also difficult to control with insecticides, in part because they feed internally. Consequently, farmer insecticide use against stem borers is generally low (Heong et al., 1994).

The food web in tropical rice fields is rich and complex; that in the Philippines includes 645 taxa and 9000 trophic (consumer resource) links (Cohen et al., 1994). Cultivation of Bt rice will result in large reductions in the abundance of stem boring and foliage feeding Lepidoptera. It is important to evaluate how this will affect generalist predators that feed on Lepidoptera as well as on other pests such as planthoppers, which will not be controlled by Bt rice. Because field testing of Bt rice is not yet permitted in tropical Asian countries, in experiments at IRRI we are using Bt sprays to simulate possible effects of Bt rice on biological control.

Impact on Soils

In fields of Bt crops, as with conventional crops, crop residues such as roots and straw are plowed back into the soil. Questions thus arise as to the fate and persistence in the soil of δ-endotoxins from crop residues and possible impacts on the soil invertebrate and microbial communities. Tapp and Stotzky (1995a) found that δ-endotoxins bind to clay in soils and become resistant to microbial degradation. Toxins bound to clay also retain their insecticidal activity (Tapp and Stotzky, 1995b). However, no changes that appeared related to δ-endotoxins were detected in microbial populations in soil samples treated with crop residues from Bt and conventional cotton (Donegan et al., 1995). Leaf extracts from transgenic maize leaves producing the CryIAb toxin were not toxic to a species of earthworm (Ciba-Geigy, 1994). However, at high doses, these extracts were toxic to a species of Collembola, small insects that are important in recycling organic matter in the soil. This finding suggests that populations of Collembola should be monitored in fields of Bt crops. In a study conducted in Oregon, Collembola populations were higher in plots of Bt potatoes (producing the CryIIIA toxin) than in plots of conventional potatoes (Monsanto, 1994).

These studies on the impact of δ—endotoxins from transgenic crop residues suggest that there will be no short-term, acute effects on soil health. However, monitoring will be necessary to evaluate long-term effects. It should also be noted that the small number of studies conducted to date have been on temperate, aerated soils. No studies have yet been done on tropical soils or submerged, anaerobic soils such as those of rice paddies.

Outcrossing of δ–Endotoxin Genes to Wild Crop Relatives

The risk of new or aggravated weed problems, resulting from the weediness of transgenic crops themselves or from the outcrossing of transgenes to wild or weedy crop relatives, has become one of the most debated environmental concerns related to agricultural biotechnology (Rissler and Mellon, 1996; Raybould and Gray, 1994). Bt crops or relatives of crop species that acquire Bt genes as a result of outcrossing would be expected to increase in fitness, and thus perhaps in weediness, only if the insects susceptible to the concerned δ-endotoxin were important constraints on the plants' distribution or abundance. Huffaker et al. (1984) document the diversity of ways in which insects can regulate plant populations. The strong impact that insects may have on plant populations is perhaps best demonstrated by the successful use of insects in classical biological control of weeds.

Historically, many of the most severe insect and weed pests of crops and pasture have been accidental introductions from other continents. Classical biological control refers to the identification of natural enemies in the native range of the pest and their release into its area of introduction. An outstanding example is the release of the moth *Cactoblastis cactorum* from Argentina to control a prickly pear cactus, *Opuntia stricta*, in Australia in the 1930s (Huffaker et al., 1984). In a few years, the cactus was brought under control in 24 million hectares of heavily infested pasture. Because the insect fauna associated with wild and weedy crop relatives is generally poorly known and the impact of insect herbivory on populations of these plants even less studied, faunistic surveys and possibly ecological experiments will in some cases be necessary to evaluate risks of weediness where fertile hybrids can form between Bt crops and wild or weedy relatives.

Risks associated with outcrossing of transgenes require particular scrutiny in centers of crop origin and diversity, both with regard to the development of new weeds and the conservation of valuable germplasm for breeding programs. Genetic engineering of crop species for eventual release to farmers in centers of origin and diversity is taking place at international and national agricultural research centers. For example, the International Potato Center (CIP) in Peru (CIP, 1994), International Maize and Wheat Improvement Center (CIMMYT) in Mexico (CIMMYT, 199?), and International Rice Research Institute (IRRI) in the Philippines (IRRI, 1996) are all developing varieties transformed with δ–endotoxin genes for enhanced resistance to caterpillar pests.

Evolution of Pest Resistance to δ–Endotoxins

At least 10 species of insects have evolved resistance to Bt spore/crystal mixtures or purified δ-endotoxins, either in laboratory studies or following extensive use of Bt in grain bins or agricultural fields (Tabashnik, 1994). It can be expected that pest species will also evolve resistance to δ-endotoxins in transgenic plants. Insect resistance to δ-endotoxins is an environmental issue in the sense that Bt is among the most effective environmentally safe insecticides available, and the loss of Bt may result in increased use of less desirable products. For organic farmers, Bt sprays are in some cases the only products suitable for their needs. Pest resistance to δ-endotoxins will, of course, also cut short the possible environmental benefits of using Bt crops themselves. The evolution of Bt resistance is of particular concern with pest species that attack multiple crops. For example, *Helicoverpa armigera* is a pest on many crops in Asia, Australia, Africa and Europe, including cotton, vegetables, and pulses. The evolution of Bt resistance in this species as a result of injudicious use of Bt cotton, for example, will have negative consequences for other crops as well.

The development of strategies to slow the evolution of pest resistance to Bt crops has become a very active area of research, and numerous recent reviews of this topic are available (e.g., Tabashnik, 1994; Andow and Alstad 1996; Gould, 1996). Based on theoretical studies and experience with chemical insecticides, it is known that the maintenance of "refuges" (toxin-free spaces or periods of time) can be an effective component of resistance management strategies. Spatial refuges can be established by, for example, interspersing fields of non-Bt crop varieties among fields of Bt varieties or mixing seeds of Bt and non-Bt before sowing. Varietal rotations or stage-specific expression of Bt genes in transgenic crops can serve to establish temporal refuges. Which kind of refuge will work best is highly dependent on the biology of the individual pest and its interaction with the crop. Maintaining refuges will in most cases require farmer cooperation, for which there is little precedent in many crop systems. Farmers growing Monsanto's Bollgard® Bt cotton, commercially released in 1996, are required to maintain refuge fields of conventional cotton (Monsanto, 1995). The progress of this resistance management program will be closely watched.

Conclusions

In contrast to many chemical insecticides, the use of Bt crops should not pose any short-term threats to the environment. However, there is a need to evaluate possible indirect or long-term effects that may result from factors such as the outcrossing of Bt genes to wild or weedy relatives or the evolution of Bt resistance in crop pests. In some cases, new research methodologies are needed to evaluate these risks and develop approaches to mitigate or avoid projected negative impacts on the environment. Any risks of the cultivation of a Bt crop must be balanced against possible environmental benefits resulting from increased yields or stability and decreased use of chemical insecticides.✧

Selected Reading

Alstad, D.N., Andow, D.A., 1996. Implementing management of insect resistance to transgenic crops. AgBiotech News and Info. 8:177N–181N.

Ciba-Geigy Corporation, 1994. Petition for determination of nonregulated status of Ciba Seeds' corn genetically engineered to express the CryIA(b) protein from *Bacillus thuringiensis* subspecies *kurstaki*. Filed with Animal and Plant Heath Inspection Service, United States Department of Agriculture.

Cohen, J.E., Schoenly, K., Heong, K.L., Justo, H., Arida, G., Barrion, A.T., Litsinger, J.A., 1994. A food web approach to evaluating the effect of insecticide spraying on insect pest population dynamics in a Philippine irrigated rice ecosystem. J. Appl. Ecol. 31:747–763.

Donegan, K.K., Palm, C.J., Fieland, V.J., Porteous, L.A., Ganio, L.M., Schaller, D.L., Bucao, L.Q., Seidler, R.J., 1995. Changes in levels, species and DNA fingerprints of soil microorganisms associated with cotton expressing the *Bacillus thuringiensis* var. *kurstaki* endotoxin. Appl. Soil Ecol. 2:111–124.

Flexner, J.L., Lighthart, B., Croft, B.A., 1986. The effects of microbial pesticides on non-target, beneficial arthropods. Agric. Ecosys. Environ. 16:203–254.

Gould, F., 1996. Deploying pesticidal engineered crops in developing countries, p. 264–293. In: G.J. Persley (ed.), Biotechnology and integrated pest management. CAB International, Wallingford, UK.

Heong, K.L., Escalada, M.M., Mai, V., 1994. An analysis of insecticide use in rice: case studies in the Philippines and Vietnam. Int. J. Pest Manage. 40:173–178.

Hokkanen, H.M.T., Wearning, C.H., 1994. The safe and rational deployment of *Bacillus thuringiensis* genes in crop plants: conclusions and recommendations of OECD workshop on ecological implications of transgenic plants containing Bt toxin genes. Biocontrol Sci. Tech. 4:399–403.

Huffaker, C.B., Dahlsten, D.L., Janzen, D.H., Kennedy, G.G., 1984. Insect influences in the regulation of plant populations and communities, p. 659–691. In: C.B. Huffaker and R.L. Raab (eds.), Ecological Entomology. John Wiley & Sons, New York.

International Maize and Wheat Improvement Center (CIMMYT), 1995. CIMMYT in 1994. CIMMY, Texcoco, Mexico.

International Potato Center (CIP), 1995. Program report for 1993–1994. CIP, Lima.

International Rice Research Institute (IRRI), 1996. Program Report for 1995. IRRI, Manila.

Jepson, P.C., Croft, B.A., Pratt, G.E., 1994. Test systems to determine the ecological risks posed by toxin release from *Bacillus thuringiensis* genes in crop plants. Molec. Ecol. 3:81–89.

Meadows, M.P., 1993. *Bacillus thuringiensis* in the environment: ecology and risk assessment. In: P.F. Entwistle et al.(ed.), *Bacillus thuringiensis*, An environmental biopesticide: theory and practice, Chapter 9, p. 193-220. John Wiley & Son, Chichester, UK.

Monsanto Corporation, 1994. Petition for determination of nonregulated status: potatoes producing the Colorado potato beetle control protein of *Bacillus thuringiensis* subsp. *tenebrionis*. Filed with Animal and Plant Health Inspection Service, United States Department of Agriculture.

Monsanto Corporation, 1995. Grower guide: essential information for growers of Bollgard®. 5 pp.

Raybould, A.F., Gray, A.J., 1994. Will hybrids of genetically modified crops invade natural communities? Trends Ecol. Evol. 9:85–89.

Rissler, J., Mellon, M., 1996. The ecological risks of engineered crops. The MIT Press, Cambridge, MA.

Schnepf, H.E., 1995. *Bacillus thuringiensis* toxins: regulation, activities and structural diversity. Curr. Opinion Biotech. 6: 305–312.

Siegel, J.P., Shadduck, J.A., 1989. Safety of microbial insecticides to vertebrates Humans. Chapter 8, p. 101–113. In: M. Laird, et al. (ed.), Safety of microbial insecticides CRC Press, Florida.

Stewart, C.N. Jr., Adang, M.J., All, J.N., Boerma, H.R., Cardineau, G., Tucker, D., Parrott, W.A., 1996. Genetic transformation, recovery, and characterization of fertile soybean transgenic for a synthetic *Bacillus thuringiensis cryIAc* gene. Plant Physiol. 112:121–129.

Tabashnik, B.E., 1994. Evolution of resistance to *Bacillus thuringiensis*. Annu. Rev. Entomol. 39:47–79.

Tapp, H., Stotzky, G., 1995a. Dot blot enzyme-linked immunosorbent assay for monitoring the fate of insecticidal toxins from Bacillus thuringiensis in soil. Appl. Environ. Microbiol. 61:602–609.

Tapp, H., Stotzky, G., 1995b. Insecticidal activity of the toxins from *Bacillus thuringiensis* subspecies *kurstaki* and *tenebriones* adsorbed and bound on pure and soil clays. Appl. Environ. Microbiol. 61:1786–1790.

Wünn, J., Kloti, A., Burkhardt, P.K., Gadab, C., Biswas, G., Launis, K., Iglesias, V.A., Potrykus, I., 1996. Transgenic indica rice breeding line IR58 expressing a synthetic *cryIA(b)* gene from *Bacillus thuringiensis* provides effective insect pest control. Bio/technology 14:171–176.

Biosafety Issues of Genetically Engineered Herbicide-Tolerant Plants—Agriculture and Agri-Food Canada's Perspective

Stephen Yarrow
Biotechnology Strategies and Coordination Office, Agriculture and Agri-Food Canada, Ottawa, Canada

Introduction

Agriculture and Agri-Food Canada (AAFC) began regulating the environmental release of plants with novel traits[†] (PNTs) in 1988. Since then, more than 3,000 confined field trials have been evaluated and authorized. Then, in 1995 the first reviews for the unconfined environmental release of some of these materials were conducted (including in most instances assessments for their use as livestock feed.) So far, there have been seventeen "precommercialization" unconfined release authorizations. Fourteen of the seventeen have been for crop plants with novel tolerances to broad spectrum herbicides. These novel herbicide tolerance traits, though not essential for the plant's survival, allow the crops to be sprayed for broad spectrum weed control during early stages of growth, when weeds can be most damaging as competitors.

Table 1 summarizes the crops and herbicides and whether the methods for introducing the novel traits were through recombinant DNA techniques, mutagenesis or somaclonal variation. Safety assessment of PNTs for human food use—another biosafety issue—is the responsibility of Health Canada, that has approved several of the plants listed in Table 1.

The assessment for environmental safety of PNTs for unconfined environmental release follows evaluation of a thorough characterization supplied by the applicant. This assessment includes consideration of five main criteria, as follows:

- Potential of the PNT to become a weed of agriculture or be invasive of natural habitats
- Potential for gene-flow to wild relatives whose hybrid offspring may become more weedy or more invasive
- Potential for the PNT to become a plant pest
- Potential impact of the PNT or its gene products on nontarget species, including humans
- Potential impact on biodiversity

Table 1. Plants with novel tolerances to herbicides: summary unconfined release authorizations (crops, herbicides, and introduction methods)

HERBICIDE	RAPESEED (canola) (*Brassica napus*)	CORN (*Zea mays*)	SOYBEAN (*Glycine max*)	FLAX (*Linum usitatissimum*)
Imidazolinone (Pursuit™)	1(1)	1(1) 1(2)	—	—
Sulfonylurea (Glean™)	—	—	—	1
Glyphosate (Round-up™)	2	—	1	—
Glufosinate ammonium (Liberty™)	3	3	—	—
Sethoxydim (Poast™)	—	1(1)	—	—

All the above are derived through recombinant DNA techniques, except (1) and (2), which were derived through mutagenesis and somaclonal variation, respectively.

In particular, plants with novel tolerances to herbicides have generated a significant amount of discussion in agricultural biotechnology and "environmental" communities in relation to the first three criteria above. (The third criterion, a PNT's pest status, relates to whether the plant could produce growth inhibiting or toxic substances or have altered adaptations to stress—characteristics that do not necessarily make a plant a weed.) Included in this considerable debate is what constitutes weediness or invasiveness. Some consider a weed to be simply a plant in a place where humankind considers it should not be. These places include agricultural and nonagricultural habitats (Rissler and Mellon, 1993). Others consider a weed to be a plant that is objectionable and unwanted in ecosystems disturbed by humans (Baker, 1965). AAFC assesses whether a plant is objectionable by considering its weedy status in "managed" ecosystems. A plant that is aggressive and displaces other species in natural ecosystems is considered by AAFC as invasive. For example, wild oat (*Avena fatua*) is a very successful weed of arable agricultural lands in Canada but has not been invasive of natural habitats. On the other hand, Purple loosestrife (*Lythrum salicaria*) is invasive of natural wetlands and has caused considerable ecosystem changes in Canada but is not considered a weed. Both species have been introduced into North America, and both have characteristics that are very well suited to the environments they inhabit—one constantly disrupted and managed by humankind, the other natural.

Central to environmental safety reviews conducted by AAFC is the thorough analysis of the biology and ecology in Canada of the plant species that has been modified. Baseline knowledge of a species' centers of origin, its related species and the potential for gene introgression into these relatives, and details of the life forms with which the species interacts is essential to effective assessments. Characterization of the potential for introduced novel traits to directly confer weediness or invasiveness properties make up part of the assessments, in addition to analysis of whether the introduced genes, coding for the novel traits, could cause indirect or unpredictable pleiotrophic effects on the plants.

The Brassica Story

Introductions of novel tolerances to herbicides into oilseed rape have caused particular interest in Canada. Both *Brassica napus* and *B. rapa* (formerly known as *B. campestris*) species are grown as oilseed rape, a major crop in Canada known as canola (grown for food oil and livestock feed meal), and rapeseed (grown for industrial oils). Concerns center on the fact that there are several weed species related to both *Brassicas* with which pollen exchange can occur. This could result in the production of interspecific hybrid seed, to which novel tolerances potentially could be passed via pollen flow and hybridization. Another complicating factor in

Canada is that in certain areas (Eastern Canada) wild biotypes of *B. rapa* are considered a nuisance, although they are not noxious or prohibited noxious weeds.

The likelihood of successful hybridizations between either *B. napus* or *B. rapa* and their related weed species and the possibility of *B. napus* X *B. rapa* hybridizations have been extensively studied by a number of scientists, in particular Warwick and Black (1993) and Scheffler and Dale (1994). They report that many interspecific and intergeneric hybridizations have been cited. Most hybrids between the various related species and with *B. napus* or with *B. rapa,* however, are usually formed only after human intervention, such as ovary culture, ovule culture, embryo rescue, or even somatic protoplast fusion. Successful *B. napus* X *B. rapa* hybridizations can be achieved through hand pollination and under natural conditions when grown in close proximity in the field (as reported in Denmark recently by Mikkelsen et al., 1996).

AAFC has concluded from considerations of the biologies and ecologies of both *B. napus* and *B. rapa* in Canada that successful introgression of novel herbicide tolerance traits (or, in fact, any trait) via pollen flow into a very few related weed species under normal field conditions is possible but at low frequencies. Introgression from *B. napus* to *B. rapa* (or *vice versa*) could occur at more measurable frequencies. The potential impact of such introgression therefore needed to be evaluated with care. AAFC came to the conclusion that the likelihood of the herbicide tolerance traits predominating and surviving in these weedy populations, in the absence of a challenge from a specific herbicide, was unlikely because the trait offers no selective advantage. However, if the particular herbicide is repeatedly used in a controlled agricultural setting, introgression of the novel tolerance trait would provide a selective advantage. This could result in weed populations immune to the effects of that particular herbicide. The commercial advantage of a specific herbicide tolerant *Brassica* crop, therefore, would quickly disappear and that herbicide would be lost as a weed control tool (e.g., sulfonylurea and sethoxydim herbicides, that are nontotal herbicidal, could be lost for use in rotation crops and total herbicidal products, such as glyphosate and glufosinate ammonium, could be lost for chemical fallow production).

During review of the six *B. napus* oilseed rape lines that have so far been authorized for unconfined release in Canada, the consequences of such introgressions into related weeds were analyzed. It was concluded that problem individuals or populations that may arise could be controlled using herbicides with different chemical or mechanical means. The same applies to "volunteer weed" populations arising from residual seeds of a previous herbicide-tolerant crop. However, AAFC is mindful that as the number of herbicide-tolerant traits increase in agriculture, the number of options available to control problem weeds or volunteers using herbicides with alternative chemical activities may decrease. To this end, biotechnology companies are aware that the intentional development of crop plants with tolerances to more than one herbicide will be viewed with caution by AAFC. In addition, mechanical means of weed control can increase soil erosion problems and so may not be a viable solution in all circumstances.

A related concern with the commercialization of several PNT lines such as oilseed rape with each tolerant to a different herbicide is the possibility of volunteer or weed populations developing multiple herbicide tolerances through natural gene flow within the species. These populations could become very difficult to control. Farmers, seed producers, and the herbicide industry, however, are keenly aware of this scenario and it is their responsibility to ensure that appropriate agronomic practices are followed to minimize this possibility. As part of the authorizations for environmental release, AAFC instructs developers of PNTs to promptly inform the Department of any new weed or volunteer developments. Developers are aware that they are responsible for controlling problem plants, should they occur. The same applies to volunteers or weeds that have adapted to a particular herbicide following its overuse. A case in point are the unconfirmed reports from Australia of annual ryegrass (*Lolium rigidum*) plants that have developed resistance to the herbicide glyphosate (Monsanto's Roundup™), a herbicide reputed to be "immune" to weed resistance. Two of the Canadian oilseed rape lines with novel herbicide tolerances, are tolerant to glyphosate. One, perhaps, should not be surprised if this supposed resistance in Australia is later confirmed—farmers anywhere should adopt responsible resistance and integrated weed-management strategies to prolong the life of any new chemical weed control agent. Uninterrupted use of the same herbicide for fifteen consecutive years is simply not a wise practice.

Other Crops

For the other novel herbicide tolerant crop species that AAFC has authorized for unconfined release in Canada (i.e., corn, soybean and flax), the introgression question is inconsequential. This is because in Canada none of these species possess weedy relatives that are even remotely sexually compatible. However, other countries will need to consider their own resident related weed populations and may come to different conclusions.

Volunteer plants of these crops could be problematical in any country, as with oilseed rape in Canada, but again, these individuals or populations can be controlled using different herbicides or mechanical practices.

Biodiversity

Potential impact on biodiversity, the fifth criterion listed before, is determined during the environmental safety assessment concerning whether the PNT will have an impact on populations of unrelated organisms in unmanaged ecosystems (the fourth criterion), such as insects. This is in addition to assessing the PNT's potential for invasiveness or gene transfer to weedy relatives. The benchmark in these assessments is the current agricultural environment, prior to the introduction of biotechnology derived plants. Of course, one could argue that some of today's agricultural practices do have detrimental effects on biodiversity. Indeed the war on weeds in a farmer's field is an intentional reduction in biodiversity—the elimination of weed species from the crop—a practice that has been going on since the beginnings of agriculture. However, the environmental safety review takes into account any changes in agricultural practices resulting from the commercialization of a PNT and ensures that these changes are neither detrimental to managed agriculture nor unmanaged habitats. In the case of plants with novel tolerances to a herbicide, the safety of the specific herbicides is also considered. The respective herbicides to which the new lines are tolerant have already existed in the Canadian agricultural market for a number of years as registered nonselective herbicides. These herbicides have already passed extensive environmental safety reviews that also take biodiversity into consideration.

Socioeconomic and Ethical Issues

The Canadian assessment process does not directly address the question of the "need" for a particular plant with novel tolerances to a herbicide. Consideration of environmental safety, human health, and changes in agricultural practices defines the current extent of the Canadian government's involvement in socioeconomic and ethical issues, in its determinations of PNT environmental releases. Of course, developers of these new plant lines claim that farmers and the environment will benefit from their use. For example, broad spectrum herbicides offer the possibility of requiring a single post-emergent application of spray following the sowing of the crop with minimal or no tillage. Developers also claim that adoption of this technology will encourage the phasing out of older herbicides in favor of newer nonselective, low dose high potency chemicals that also quickly degrade in the soil. Herbicide-tolerant plants may indeed be a useful solution to the substantial losses in crop yields caused by weeds.

The role of government regulators, nonetheless, is limited to maintaining environmental and human safety (in addition to verifying efficacy claims), whatever the declarations of developers. Society as a whole will ultimately have to decide whether these new biotechnology derived crops will be beneficial to agriculture and consumers. This behoves governments to provide effective communications about what plants have been assessed and approved. To this end, AAFC provides "decision documents" on each PNT approval to anyone who is interested and has recently set up a web site for this purpose on the Internet (*http://www.agr.ca/fpi/agbiotec/commarea/plants/home.html*).

Some critics of the methods used by the Canadian regulatory system in dealing with products of biotechnology, believe that the government has an ethical obligation to enhance the sustainability of agriculture. They suggest that regulatory approvals of plants with novel tolerances to herbicides will increase and perpetuate chemical-dependency of agriculture, which conflicts with the view of sustainable agriculture. Canada's regulatory system does not require proof of positive environmental or socioeconomic benefits. Instead, the Canadian government designed the regulatory system to focus assessments on preventing potential negative impacts to the environment, to human health and on agronomic practices which effect sustainability—an ethically sound focus in itself. Agricultural sustainability remains a priority for the government of Canada.

Other critics of biotechnology are more concerned with genetic engineering itself by which technology facilitates the transfer of genes from one species to another. Although important concerns, public perception and the labeling of biotechnology-derived food products are beyond the scope of this paper.

Regulatory Triggers

Not all of the fourteen novel herbicide-tolerant crop lines authorized for unconfined release were derived through genetic engineering. Nonetheless, the fact that they were all subject to the same review criteria is a testimony to Canada's regulatory approach; i.e., regulating the product and not just products of specific processes. The development process is examined as part of safety review, but it is not the trigger for regulatory review. The potential introgression and volunteer problems, described above, apply equally to novel herbicide tolerance genes introduced through recombinant DNA techniques, mutagenesis, or somaclonal variation.

A crop trait's novelty is determined through analysis of familiarity and substantial equivalence to crop lines of the same species already in Canadian agricultural systems. Developers of herbicide-tolerant plants require regulatory approval only if their new crop line is either unfamiliar or familiar, but not substantially equivalent.

Lack of familiarity is based on whether the new line is a new species to Canada, whether the specific herbicide tolerance trait is new to Canadian agriculture (including whether the specific mechanism of tolerance is new), and if the cultivation practices of the new line will be different from that already used in Canada. The new line may be familiar, but not substantially equivalent, if the developer does not know whether it may result in altered environmental interactions compared to its counterparts. This analysis of familiarity and substantial equivalence of novel traits is really the crux of the Canadian safety-based regulatory approach.

The regulatory approach adopted by Canada is perhaps more inclusive than the approaches developed in the United States and Europe. This is because it focuses on whether new plant lines are exhibiting novel traits, whatever the method of introduction. A good example to illustrate this point, is the oilseed rape line recently authorized in Canada that has novel tolerance to imidazolinone-based herbicides. This line was derived through mutagenesis and tissue culture and not recombinant DNA techniques, but it was still subject to the above described PNTs assessments before it could be commercialized in Canada. In the U.S. and Europe, the same line was treated as a traditionally derived variety with no assessment.

Despite the process in which the imidazolinone-tolerant rapeseed was derived, the main focus of the Canadian review was on the potential spread of the tolerance trait to related weeds. On the other hand, the safety-based regulatory approach used in Canada will be flexible enough not to capture all biotechnology-derived products if they pass the familiarity and substantial equivalence analyses.

Conclusion

Plants with novel herbicide tolerances may become a significant part of Canada's agricultural system within the next few years and regulatory reviews of further plant lines are anticipated. Ultimately, the marketplace will decide how popular these lines will be to Canadian growers, compared to currently available varieties. Any negative effects on the environment and biodiversity, however, should be minimal by "passing" AAFC's rigorous regulatory reviews. How the Canadian government considers socioeconomical and ethical issues remains under review. Nevertheless, future considerations on such issues are not likely to significantly change the current regulatory process as it relates to safety assessments.

Signatories to the Convention on Biological Diversity, introduced at the 1992 "Rio Summit," are currently considering a Biosafety Protocol that will provide a worldwide regulatory mechanism for the transboundary movement of living modified organisms. Canada's regulatory approach for the environmental release of plants with novel traits could perhaps be a useful contribution for consideration at the upcoming protocol negotiations.✧

† A plant with a novel trait is defined by AAFC as a plant variety/genotype possessing characteristics that demonstrates neither familiarity nor substantial equivalence to those present in a distinct, stable population of a cultivated species in Canada and that have been intentionally selected, created or introduced into a population of that species through a specific genetic change—regardless of how the genetic change was made.

Selected Reading

Baker, H., 1965. Characteristics and modes of origin of weeds, p. 147-168. *In*: H.G. Baker and G.L. Stebbins (eds.), Genetics of colonizing species. Academic Press, New York.

Mikkelsen, T.R., Andersen, B., and Jørgensen, R.B., 1996. The risk of crop transgene spread. Nature 360:31 (March 7).

Rissler, J. and Mellon, M., 1993. Perils amidst the promise: ecological risks of transgenic crops in a global market. Union of Concerned Scientists. USA.

Scheffler, J.A. and Dale, P.J., 1994. Opportunities for gene transfer and origin of crop Brassicas, a review. Opera Bot. 55:3–57.

Warwick, S.I. and Black, L.D., 1993. Guide to the wild germplasm of Brassica and allied crops, Part III: Interspecific and intergeneric hybridization in the tribe Brassiceae (Cruciferae). Technical Bulletin 1993–16E, Centre for Land and Biological Resources Research, Agriculture and Agri-Food Canada.

Biosafety Issues of Genetically Engineered Virus-Resistant Plants

H. R. Pappu
Department of Plant Pathology, University of Georgia, College of Agriculture and Environmental Sciences, Tifton, Georgia, USA.

Abstract

Viruses constitute a major threat to global food security, reducing the yield and quality of many crops. The most effective and durable approach for the management of viral diseases is the use of virus-resistant plants. Advances in plant biotechnology have contributed toward achieving this goal, especially in cases where sources of resistance are not available or, if available, are not amenable to incorporation into cultivated crop plants. Research during the past decade has shown that incorporating defined viral sequence(s) into plants may confer resistance to challenge infection by homologous and in some instances heterologous viruses. This genetically engineered resistance is being tested for its effectiveness under field conditions in various crops in several countries.

Such field tests under natural conditions showed promising results in reducing the impact of viral infections in several crops. However, concerns have been raised about the possibility of generation of new viruses as a result of interactions between the transgene and the challenge virus. Mechanisms by which new viruses or strains could emerge include recombination, and transcapsidation. Both events have resulted in the production of recombinant virus from a transgenic plant under experimental conditions. This should be viewed in context of the frequency of such genetic event(s), and the level of selection pressure exerted in nature. Approaches to reduce the frequency of recombination and/or the effects of such events in transgenic plants expressing viral sequences include a) using un-translatable RNA sequences of viral origin so that expression of these sequences in the plant does not lead to expression of the viral gene expression; b) expressing defective/truncated forms of the viral gene; c) avoiding viral sequences that might initiate recombination (such as 5' and 3' ends of RNAs); and d) selecting transgenic plant lines that express minimal levels of the transgene. The potential risks of using virus-resistant transgenic plants should be weighed against the benefits such as increased yield, better quality, and reduced use of pesticides.

Introduction

Viral diseases are limiting factors for the production of the world's crops that provide food, feed, fiber, and oil. Viral diseases of international importance include barley yellow dwarf, rice tungro, citrus tristeza, beet yellows, tomato spotted wilt, and several diseases caused by geminiviruses and potyviruses. Losses due to viral diseases are more acute in tropical, developing countries.

The management of viral diseases depends on several factors: the identity of the causal virus, its biology, molecular biology, and epidemiology, as well as the availability of sensitive and reliable detection tools and methodologies. Such information is essential to develop control strategies in a given crop-virus combination. Viral constraints of crops can be effectively managed through a combination of several different strategies: using reliable diagnostic tools, cultural practices, management of vectors, use of virus-free planting material, prevention of the introduction of a new virus by having an effective quarantine program, and through the use of virus resistant cultivars. These approaches have been largely successful in minimizing the losses due to viral diseases in different parts of the world (Matthews, 1991).

The most effective, durable, and cost effective approach to controlling viral diseases is by using virus resistant cultivars. The use of resistant cultivars not only facilitates the realization of a crop's full yield potential, but also reduces the use of chemicals for controlling the vectors. Thus, the benefits are various and include: reduced usage of pesticides, reduced costs of production, better quality, higher yield, and increased sustainability.

While there are several examples of the successful development and use of virus resistant plants, this approach is still constrained by several factors. Genes for resistance to a particular virus may not readily be available in cultivated species, thus requiring an expensive and time consuming search for resistance in related cultivated and wild species. Even if a source of resistance is available, it may not be amenable to transfer into an existing cultivar. In spite of these limiting factors, development and use of virus resistant plants remains the most effective method for controlling virus diseases.

The Idea Behind Virus-Resistant Transgenic Plants

One approach for management of virus diseases in use for many years is the phenomenon called cross protection. First observed in nature in the 1930s, cross protection is defined as the phenomenon in which plants infected by a mild strain of a virus become resistant to infection by a severe strain of the same virus. Cross protection was successfully used to minimize losses in Citrus due to citrus tristeza closterovirus, in papaya due to papaya ring spot potyvirus, and in greenhouse-grown tomatoes due to tomato mosaic tobamovirus. The basis for cross protection is thought to result from an environment that is not conducive for the multiplication of a more aggressive virus strain in a plant cell that is already infected by the mild strain. The ability of a mild strain to confer resistance seems to be highly specific, and is largely limited to interactions between strains of the same virus. It is a well-known fact that not all strains of a given virus have the ability to cross protect. Careful screening and selection of mild, cross-protecting strains remain the key to successful management of viral diseases using this strategy (Rocha Pena et al., 1995).

It has been theorized that this antiviral state can be induced by genetic manipulation (Hamilton, 1980; Sanford and Johnson, 1985). Advances in plant molecular biology, tissue culture, and microbial molecular biology in the last two decades have provided the tools to explore the possibility of genetically engineered resistance for virus control. Improved protocols for regeneration of plants, development of transformation systems based on *Agrobacterium*, discovery of constitutive promoters such as 35S from cauliflower mosaic caulimovirus, and other methods for gene delivery such as the "gene gun," increased information on the structure and genome organization of several plant viruses have contributed toward achieving this task.

Virus-Resistant Transgenic Plants: The Breakthrough

The first successful demonstration of genetically engineered cross protection against a plant virus was reported in 1986 by Dr. Roger Beachy and his group,

Table 1. List of virus-resistant transgenic plants field-tested in the U.S. (1989–1996)

Crop	Gene	Source of Gene
Alfalfa	Coat protein	Alfalfa mosaic
Barley	Coat protein	Barley yellow dwarf
Beet	Coat protein	Beet necrotic yellow vein
Corn	Coat protein	Maize chlorotic mottle
	Coat protein	Maize chlorotic dwarf
	Coat protein	Maize dwarf mosaic
Cucumber	Coat protein	Cucumber mosaic
	Coat protein	Cucumber mosaic, watermelon mosaic 2, zucchini yellow mosaic
	Coat protein	Papaya ring spot, squash mosaic, Papaya ring spot
	Nuclear incl. protein A	Papaya ring spot
	Nuclear incl. protein B	
Gladiolus	Coat protein	Bean yellow mosaic
Lettuce	Nucleocapsid	Tomato spotted wilt
Melon	Antisense coat protein	Zucchini yellow mosaic
	Coat protein	Cucumber mosaic, papaya ring spot,watermelon mosaic2
	Coat protein	Zucchini yellow mosaic
	Nuclear incl. protein A	Watermelon mosaic 2 and zucchini yellow mosaic
		Papaya ring spot
	Nuclear incl. protein B	Papaya ring spot
Papaya	Coat protein	Papaya ring spot
Peanut	Coat protein	Tomato spotted wilt
	Nucleocapsid	Tomato spotted wilt
Pepper	Coat protein	Cucumber mosaic

Crop	Gene	Source of Gene
Potato	Coat protein	Bean yellow mosaic
	Coat protein	Potato leaf roll
	Coat protein	Potato virus X
	Coat protein	Potato virus Y
	Coat protein	Tobacco vein mottling
	Coat protein	Barley yellow dwarf
	Coat protein	Tobacco rattle
	Coat protein	Tobacco vein mottling
	Coat protein	Potato virus X and potato virus Y
	Coat protein	Potato virus X, potato virus Y, and potato leaf roll
	Coat protein	Potato virus Y and potato leaf roll
	Antisense coat protein	Potato leaf roll
	Antisense coat protein	Potato virus Y
	Antisense coat protein	Tobacco rattle
	Antisense coat protein	Barley yellow dwarf
	Antisense coat protein	Tobacco vein mottling
	17 Kda	Barley yellow dwarf
	17 Kda	Potato leaf roll
	34 Kda	Tobacco vein mottling
	42 Kda	Tobacco vein mottling
	60 Kda	Tobacco rattle
	Vpg	Potato leaf roll
	Vpg	Potato virus Y
	Replicase	Potato leaf roll
	Protease	Potato leaf roll
	Helper component	Tobacco vein mottling
	Nuclear incl. protein A	Potato virus Y
	Nuclear incl. protein B	Potato virus Y
Plum	Coat protein	Papaya ring spot
	Coat protein	Plum pox
Soybean	Coat protein	Soybean mosaic
	Coat protein	Bean pod mottle

Crop	Gene	Source of Gene
Squash	Coat protein	Papaya ring spot
	Coat protein	Cucumber mosaic
	Coat protein	Cucumber mosaic and papaya ring spot
	Coat protein	Watermelon mosaic 2 and zucchini yellow mosaic
	Coat protein	Cucumber mosaic virus, papaya ring spot, watermelon mosaic 2,
	Nuclear incl. protein A	Squash mosaic, and zucchini yellow mosaic, Papaya ring spot
	Nuclear incl. protein B	Papaya ring spot
Tobacco	Antisense coat protein	Tobacco etch
	Coat protein	Tobacco etch
	Coat protein	Potato virus Y
	Coat protein	Alfalfa mosaic
	Coat protein	Beet curly top
	Coat protein	Tobacco vein mottling
	Cylindrical inclusion	Tobacco vein mottling
	Helper component	Tobacco vein mottling
	2–5A dependent RNase	Human
	2–5A dependent RNase	Human
	2–5A synthetase	Human
	Protein kinase	Human
	Replicase antisense with ribozyme	Tobacco mosaic
Tomato	Nucleocapsid	Tomato spotted wilt
	Coat protein	Tomato yellow leaf curl
	Coat protein	Tomato mosaic
	Coat protein	Cucumber mosaic
	Replicase	Tobacco mosaic
	Satellite CARNA 5	Cucumber mosaic
	C1 Antisense	Tomato yellow leaf curl
	C2 Antisense	Tomato yellow leaf curl
	C3 Antisense	Tomato yellow leaf curl
	Movement protein	Tobacco mosaic virus
Watermelon	Coat protein	Watermelon mosaic 2 and zucchini yellow mosaic
Wheat	Coat protein	Wheat streak mosaic
	Coat protein	Barley yellow dwarf

then at Washington University in St. Louis, MO, USA (reviewed by Fitchen and Beachy, 1993). They showed that tobacco transformed with the capsid protein gene of tobacco mosaic virus (TMV) resisted further infection by TMV. This resistance was expressed as a delay in the development of the disease (symptoms). While the plants were resistant to an inoculation with a virion preparation, they were found to be as susceptible as nontransformed control plants when inoculated with viral RNA. It was hypothesized that the resistance was due to blocking of an early event in virus replication such as uncoating of the virion RNA upon entry into the plant cell. This successful demonstration of genetically engineered cross protection was quickly followed by similar reports for other viruses.

While tobacco remained as the host of choice, several other crops of economic importance were also used to test the effectiveness of this management strategy (reviewed by Pappu et al., 1995). This technology has now progressed from being tested in the laboratory/greenhouse to the field. A wide range of virus resistant transgenic plants have been field tested in the United States (U.S.) in the last few years (Table 1). Of the various strategies used to achieve resistance, the most common one was the expression of capsid protein gene sequence. Other approaches include expressing other viral genes such as replicase (complete or parts thereof), protease, Vpg, and nuclear inclusion body proteins (Table 1). Another approach found to be effective in conferring virus resistance was the use of viral gene sequences that were rendered untranslatable (by introducing mutations such as stop codons that interrupted the translation). Viral satellites with the ability to ameliorate the symptoms of their respective helper viruses also are being used in transgenic plants to engineer resistance to certain plant viruses (Yie and Tien, 1993). Several reviews provide detailed information on many aspects of virus resistant transgenic plants (Fitchen and Beachy, 1993; Pappu et al., 1995; Lomonossoff, 1995).

Concerns About Using Virus-Resistant Transgenic Plants

Concerns were raised by deZoeten (1991) regarding the risks associated with using virus-resistant transgenic plants. The greatest anticipated risk was the potential for generation of new viruses, and subsequently new virus diseases. The following section will describe the potential risks. The Proceedings of a recent workshop offers an excellent source of information on this topic (American Institute of Biological Sciences, 1995).

Let us first consider the ways in which new viruses or new virus variants may arise. During the course of evolution, the present day diversity of viruses may have arisen due to the inherent ability of viruses to change. This change is believed to be due to the high error rate of RNA polymerases leading to variants that may survive and eventually get established as a new entity.

It is a currently widely accepted theory that present day viruses were derived through this evolutionary process. Another theory put forward and based on sequence and genome organization comparisons is that the "starting point" of viruses was a basic set of gene blocks that became rearranged in different combinations and were further influenced by various levels of selection during the course of evolution.

There are three phenomena that provoke the recent concern regarding the risks associated with the use of virus-resistant transgenic plants: 1) phenotypic mixing, 2) synergism, and 3) genetic recombination. Most of the scientific community in the United States agrees that the use of virus-resistant transgenic plants does not pose any greater risk than that encountered through the use of "traditional" resistant varieties. Also, there is a general agreement that monitoring transgenic plants on a large scale is impractical, if not impossible.

Consensus was reached on the need to carefully evaluate and estimate the risk of recombination leading to generation of new viruses. Several research projects are underway to evaluate and quantitate some of these risks.

Phenotypic Mixing

Phenotypic mixing is defined as the encapsidation of the RNA of virus A by the capsid protein of virus B. This results in cases of mixed infections where a plant is infected by two viruses. The RNA of virus A will then "acquire" the properties of the capsid protein of virus B.

Technically, a "new" virus will have emerged due to this process. However, the new properties acquired by the new virus are limited to only one generation, since the RNA of virus A will subsequently be encapsidated by its own capsid protein.

Mixed virus infections are extremely common in nature, especially in perennial crops, such as apple and citrus, as well as many annual crops. However, there is no evidence of generation of a viable and persistent new virus as a result of mixed infection, probably due to the fact that these hybrids tend to be very unstable and do not survive in nature.

Synergism

Synergism can be considered as the opposite of cross protection. While cross protection is described as the inability of two viruses or two virus strains to multiply in the same host; synergism is the result of more than the additive effect of infections by two viruses. Infection of a plant by two viruses may create a more severe disease than that caused by either of the viruses alone. The most celebrated example of synergism in nature is the mixed infection of potato virus X and potato virus Y (Matthews, 1991). Even in transgenic plants, expression of certain genes from one of the synergistic "pair" can result in a synergistic reaction upon infection by the other member of the pair. Since not much is known about the basis for the synergistic effect, more work is needed to understand the molecular mechanism(s) that influence this interaction. For now, the risk posed by viral synergism in transgenic plants can be dealt with by screening the crop and transgene in question against potential viruses that might create the synergetic effect.

Genetic Recombination

Recombination or the exchange of genetic material between organisms is one of the major avenues through which biological diversity generated. There are two types of recombination: homologous and heterologous. Genetic recombination has been studied in great detail in eukaryotic organisms. Little was known about the occurrence of recombination in RNA viruses of plants until a few years ago. Due to the development of infectious RNA transcripts, it is now possible to study structure-function relationships with respect to recombination in plant viruses. Nevertheless, present day genetic diversity among plant viruses may be due to recombination over the course of evolution. This may be attributed to the relative high error rate of RNA polymerases involved in viral replication. The most widely accepted mechanism of recombination in plant viruses is through copy choice mechanism. It is described as the tendency of the RNA polymerase to switch templates. This may take place between closely related RNA templates (homologous recombination) or divergent templates (heterologous recombination). In recent years several reports were published showing that recombination does take place among plant viruses (reviewed by Simon and Bujarski, 1994). Whether the use of virus-resistant plants will increase the risk of recombination is difficult to assess at this time, since it should be compared with natural occurrence of such recombinational events. However, little or no information is available on the frequency of natural recombination among plant viruses. In view of the lack of data, it is especially difficult to evaluate the risk of recombination created by the use of virus resistant transgenic plants.

The first experimental evidence that a transgene can recombine with an infecting virus from which the transgene was derived, was provided by Greene and Allison (1994). This confirmed the earlier speculation about the role of recombination in generating new strains or viruses. However, this should be viewed in the context of the selection pressure to which the transgene and the infecting virus was subjected. Is the selection pressure applied in controlled experiments indicative of the one that is encountered in nature? If not, is it possible to simulate the natural selection pressure under laboratory conditions? If so, the first task would be to assess the level of selection pressure in nature, which is difficult, if not impossible. Before one can estimate the risk of recombination in generating new viruses, one needs to determine the rate of recombination between a transgene and an infecting virus at different levels of selection pressure and in different environments.

Other aspects pertaining to the rate of recombination among plant viruses that need to be considered are whether the rate of recombination is more likely to occur within a virus group compared to members of different virus groups (taxa). To date, there is no evidence of recombination between viral groups (taxa). However, evidence indicates that recombination takes place under experimental conditions between members of the same taxa; e.g., within the bromovirus group (Simon and Bujarski, 1994). Such recombination may occur in nature as well, but the resulting recombinants may go undetected because most are not viable.

While phenotypic mixing and synergism may not be major contributing factors for generating new viruses through the use of virus-resistant transgenic plants, genetic recombination, however, should be carefully evaluated for its ability to contribute to this risk. For a recombinant "new" strain or virus to become viable and "successful" as a plant pathogen, it must replicate, it must be transmitted to other plants, it must invade/infect other plants upon transmission, and its progeny must survive and be able to continue the infection cycle.

Does the wide-scale use of virus-resistant transgenic plants increase the frequency of recombination, thereby increasing the odds of the above events? One scenario would see an increase in the frequency. Another scenario could predict this since virus-resistant plants suppress the replication of the infecting virus. In this case, there will be relatively fewer molecules of the infecting viral genome to recombine with the transgene. Thus, the frequency of recombination may not be any higher than that seen in naturally occurring mixed virus infections. The chances of all these events occurring to make a new virus "successful" are minimal.

To minimize the risk of genetic recombination even further, certain strategies can be adopted. They are as follows:

- Avoiding 5' and 3' ends of RNAs as part of the transgene since these sequences are believed to increase the chances of recombination.
- Using nontranslatable viral sequences as transgenes so that no gene product is produced from the transgene.
- Introducing more stop codons in their sequences so that there would be further reduction in the expression of the transgenes.
- Selecting transgenic lines that express minimal levels of transgene so that the concentration of the cellular transgene is below the limit that would permit recombination between the transgene and the infecting viral genome.
- Using sequences that code for part of the replicase gene.
- Avoiding use of satellite RNAs with a potential to cause more severe disease (Palukaitis and Roosinck, 1996).

In addition to these concerns mentioned, the use of transgenic plants, in general, has drawn attention with respect to two other widely publicized potential risks. Will the transgene flow into the crop's wild relatives; i.e., through pollen? If so, will such an "escape" increase the weediness?

Little or no data is available to evaluate these risks posed by the use of virus-resistant transgenic plants. In theory, if the transgene escapes into a wild relative, the wild relative should be resistant to further infection, leading to a reduction in the virus inoculum, since that particular weed would not be able to serve as a virus reservoir.

Fuchs and Gonsalves (1996) reported that experiments are being conducted to estimate the extent of such gene flow into *Cucurbita texana*, a wild relative of squash (*C. pepo*) from transgenic squash expressing the CP genes of zucchini yellow mosaic potyvirus (ZYMV), cucumber mosaic cucumovirus (CMV), and watermelon mosaic 2 potyvirus (WMV–2). The risk of the transgene flow into *C. texana* and the ability of the F1 hybrids of *C. texana* x transgenic squash are being studied under field conditions (Fuchs and Gonsalves, 1996).

Use of Virus-Resistant Transgenic Plants: Current Status

The last few years have seen significant activity to utilize this technology in commercial products. Several plant biotechnology companies in the U.S. and Europe are in the forefront of this. Following the approval by USDA-APHIS, the first commercially available product as a result of the use of a virus-resistant transgenic plant was introduced in the U.S. by Asgrow. Transgenic squash resistant to three viruses (CMV, ZYMV, and WMV2) is now sold in supermarkets in selected states in the U.S. Risk assessment studies by USDA-APHIS concluded that virus-resistant transgenic squash does not pose a greater risk than nontransgenic crops with respect to generating new viruses. Transgenic papaya, resistant to papaya ring spot virus, was recently approved for commercialization by USDA-APHIS. Applications for similar commercialization are judged on an individual case-by-case basis by USDA-APHIS, it can be expected that more products derived from virus-resistant transgenic plants are likely to obtain regulatory approval for commercialization in the very near future.

The most important questions one must answer in making a judgement whether or not to deploy virus-resistant transgenic plants for managing virus diseases include the following:

- Does the use of virus-resistant transgenic plants pose greater risks than those developed through traditional plant breeding?
- Do transgenic plants exert greater selection pressures for the generation of new viruses than those exerted by use of traditional varieties?

The paucity of information available on the role of traditional varieties in generating resistance-breaking strains of the virus makes it impossible to evaluate the risk for transgenic plants without more research in this area.

Living With Risks

Monitoring or evaluating potential risks may be possible on a small scale, but it is impractical on a large scale. Risks also should be viewed in the context of the need to control the damage caused by viruses. This need for control acquires additional significance in the case of developing countries where losses to viruses can reach as high as 30%. Minimizing the losses due to pests and diseases, in general, would go a long way toward improving the production of several crops.

While increasing the crop productivity should continue to be a top priority to meet the ever growing demands of human population, equal emphasis should be given to realizing the existing yield potential through management of biotic stresses, such as those caused by viruses. Virus-resistance through transgenic technology offers an immediate and effective management strategy that should be considered as another tool in achieving that goal. A major potential contribution of transgenic plant technology is the relative ease with which multiple virus resistance genes can be introduced into crops. Thus, virus-resistant transgenic plants offer a significant advantage over those developed by traditional breeding methods.

The ultimate decision to deploy this technology rests on individual regions/countries and their assessment of real and perceived risks. In spite of these risks, virus-resistant transgenic plants will continue to be explored for their ultimate use in managing virus diseases. In the U.S., with the recent approval by USDA-APHIS of transgenic squash, one can expect to see more products reaching the market in the near future. It remains to be seen how this technology will be accepted by and made available to the developing countries.✧

Selected Reading

American Institute of Biological Sciences, 1995. Transgenic virus resistant plants and new viruses. Proceedings of a workshop, Beltsville, MD. April 20–21, 1995.

deZoeten, G.A. 1991. Risk assessment: do we let history repeat itself? Phytopathology 81:585–586.

Fitchen, J.H. and Beachy, R.N., 1993. Genetically engineered protection against viruses in transgenic plants. Annual Review of Microbiology 47:739–763.

Fuchs, M., and Gonsalves, D., 1996. Is gene flow a serious risk for the release of virus-resistant transgenic crops? Phytopathology 86, No. 11 (Supplement):S43.

Greene, A.E., and Allison, R.F., 1994. Recombination between viral RNA and transgenic plant transcripts. Science 263:1423–1425.

Hamilton, R.I., 1980. Defenses triggered by previous invaders: Viruses, p. 279–303. In: J.G. Horsfall and E.B. Cowling (eds.), Plant disease: an advanced treatise, Vol. 5. Academic Press, New York.

Hull, R., 1995. Release of viral transgenic plants to the environment: prospects and problems. Mitt.a.d. Biol. Bundesanst.H. 309:17–22.

Falk, B.W. and G. Bruening, 1994. Will transgenic crops generate new viruses and new diseases? Science 263:1395–1396.

Kling, J., 1996. Could transgenic supercrops one day breed superweeds? Science 274:180–181.

Lomonossoff, G.P., 1995. Pathogen-derived resistance to plant viruses. Annual Review of Phytopathology 32:337–343.

Matthews, R.E.F., 1991. Plant virology. 3rd Edition. Academic Press, New York.

Palukaitis, P., and M.J. Roossinck, 1996. Spontaneous change of a benign satellite RNA of cucumber mosaic virus to a pathogenic variant. Nature Biotechnology 14:1264–1268.

Pappu, H.R., C.L. Niblett, and R.F. Lee, 1995. Application of recombinant DNA technology in plant protection: molecular approaches to engineering virus resistance in crop plants. World Journal of Microbiology and Biotechnology 11:426–437.

Rocha-Pena, M., R.F. Lee, R. Lastra, C.L. Niblett, F. Ochoa-Corona, S.M. Garnsey, and R.K. Yokomi, 1995. Citrus tristeza virus and its aphid vector Toxoptera citricida: serious threats to citrus production in the Caribbean countries and Central and North America. Plant Disease 79:437–445.

Sanford, J.C., and S.A. Johnson, 1985. The concept of parasite-derived resistance: deriving resistance genes from the parasite's own genome. Journal of Theoretical Biology 113:395–405.

Simon, A.E., and J.J. Bujarski, 1994. RNA-RNA recombination and evolution in virus-infected plants. Annual Review of Phytopathology 32:337–362.

Yie, Y. and Tien, P., 1993. Plant virus satellite RNAs and their role in engineering resistance to virus diseases. Seminars in Virology 4:363–368.

Marker Gene Controversy in Transgenic Plants

Vedpal S. Malik

U.S. Department of Agriculture, Animal and Plant Health Inspection Service, Riverdale, Maryland, USA

Introduction

Selectable markers are genes that allow cells expressing them to be descriminated selectively out of large populations. Traditional breeders identify observable markers that are linked to traits of interest. Recombinant DNA is used to physically link marker genes to DNA sequences in one piece that confer the desired phenotype on the plant. This is achieved by inserting the DNA sequences for the desired trait into vector molecules that carry a selectable marker gene. A vector is a DNA molecule that is used as a vehicle to carry foreign DNA sequences into *E. coli* or other host cells. In certain instances, the marker gene is also the gene of interest (e.g., herbicide resistance).

Plasmids are the simplest vectors. They are circular DNA molecules that exist separately from the main chromosome. To be propagated through successive bacterial generations, the self replicating plasmid vector contains specific DNA sequences that allow its replication in the host cells. DNA polymerase and other proteins required for initiation of DNA synthesis bind to this region called the origin of replication (ori). The vectors that are used for introducing genes in plants have *E. coli* compatible origin of replication and do not autonomously replicate in the plant. They get stably integrated in the plant chromosomes and replicate as any other chromosomal genes.

The gene of interest is first cloned in a vector containing a selectable marker gene. After transformation, the marker gene is used to select rare transformed cells that contain the added linked trait from large numbers of nontransformed cells. The selectable marker gene allows the growth of cells expressing the marker gene on a selective media while eliminating the cells without the marker gene. The frequency with which transformed cells are obtained are often as low as 1 in 10,000 and finding these rare events without selectable markers would be very difficult. Selectable markers have their own promoters and terminators that are often host specific. The type of promoter dictates whether a selectable marker gene will be expressed in *E. coli* or plants.

Two types of selectable marker gene sequences can end up in transgenic plants: (1) the selectable marker gene sequences that are used for initial cloning and manipulation of genes in *E. coli* and (2) the selectable marker gene sequences that are expressed in plants and are used to select transformed plant cells. In some vectors, kanamycin selectable marker is engineered so that it is expressed in both plants and bacteria. Several categories of selectable markers genes are given in Table 1.

Marker genes for monitoring genetically modified microorganisms released into nature must be stable in the host and may be integrated into the chromosome of the released microorganisms. Marker genes must be detectable in the natural environment at very low frequencies. The marker gene should not produce a toxic metabolite that could harm threatened and endangered species. It should not increase the pathogenicity of the host microorganism.

The marker gene must not be deterimental to the growth and fitness of the recipient microorganism so that the information obtained using it reflects the behavior of the unmodified parent in the environment. Auxotropic markers often make the recipient microorganism less fit. Marker genes that have a positive effect on the survival of the recipient microorganisms are desirable. Marker genes conferring resistance to antibiotics of great clinical significance, particularly those to which resistance is not widely spread in the environment, should be avoided.

Selectable Markers for Escherichia coli

Selectable markers in *E. coli* allow selection of only those cells which contain a vector expressing a selectable marker. Such markers are of two types: drug resistance and auxotrophic. A drug resistance marker enables cells to detoxify an exogenously added drug that would otherwise kill the cell. Auxotrophic markers allow cells to synthesize an essential component (usually an amino acid) in media which lacks that essential component. Drug resistance markers in most popular *E. coli* plasmid vectors encodes for resistance to antibiotics like tetracycline, kanamycin, streptomycin, zeocin and ampicillin.

Table 1. Selectable markers genes expressed in transgenic plants field tested in the United States

Marker Gene	Donor Organism	Regulated Article	Applicant	APHIS Reference
Acetolactate synthase	Tobacco	Cotton	DuPont	95–256–01p
β-glucuronidase	*E. coli*	Apple	Cornell University	95–088–01r
5-enolpyrvylshikimate-3-phosphate synthase	*E. coli, Agrobacterium tumefaciens*	Beet	American Crystal Sugar Company	96–099–01r
Hygromycin phosphotransferase	*E. coli*	Rapeseed	University of Georgia	94–200–01r
Glyphosate oxidoreductase	*Achromobacter* sp	Corn	Wyffels Hybrid, Inc.	96–142–02n
Phosphinothricin acetyl transferase	*Streptomyces hygroscopicus*	Corn	Ciba Seeds	94–319–01p
Nitrilase	*Klebsiella ozanenae*	Cotton	Calgene, Inc.	93–034–01r
β-glucuronidase	*E. coli*	Soybean	AgrEvo	96–068–01p
Neomycin phosphotransferase	*E. coli*	Tomato	Agritope	96–068–01p
Neomycin phosphotransferase	*E. coli*	Tomato	DNA Plant Technology	94–228–01p
Luciferase	*Photinus pyralis*	Corn	DuPont	93–077–01r
Chloramphenicol acetyltransferase	*E. coli*	Potato	Washington State University	92–325–01r

In a routine cloning experiment a plasmid vector containing a β-lactamase gene (Amp) is first cleaved with restriction enzymes and mixed with the gene that is targeted for cloning. After linking the vector and the target gene with the enzyme ligase, the newly created hybrid and parent molecules are introduced into *E. coli* by the process called transformation. The transformed *E. coli* cells are then grown on agar media that contains a selecting agent; i.e., ampicillin. Only the individual cells that acquire the selectable marker gene on the vector survive, multiply, and develop into a clonal colony. This is so because the selectable marker gene codes for an enzyme such as β-lactamase that inactivates the selecting agent, ampicillin, in the medium. Cells without the selectable marker gene die on the selective media. In this way antibiotic resistance is a selectable marker that allows identification of cells that have taken up plasmid vector DNA, including recombinant plasmid that contains the gene of interest. Appropriate recombinant plasmid with desired gene combinations which is a rare event is identified and introduced into plant cells by a number of methods. The transformed plant cells are now selected using a selectable marker gene that is functional in plant cells; e.g. herbicide resistance.

Beyond its use for eliminating the large number of nontransformed *E. coli* or plant cells, a selectable marker like Amp or glyphosate-tolerance is generally not necessary for expression of the accompanying traits engineered for expression in plants; i.e., selectable markers for antibiotic resistance used in *E. Coli* and glyphosate-tolerance used in plants are not necessary for the engineered novel trait; e.g., pesticidal function of the plant. In the absence of selection pressure, selectable markers may even be lost from the plant during subsequent cultivation with no effect on the desired engineered plant phenotype. Vectors are now available that allow further distinction between clones that acquired recombinant, nonrecombinant, and parental vectors. Examples of this are insertional inactivation of the β-galactosidase gene which when undisrupted produces a blue reaction product from a chromogenic substrate. Insertion of the target DNA into the β-galactosidase gene flags the transformed clones with the recombinant plasmid. When plated on appropriate media, the wild type cells produce blue colonies while recombinant clones are white in color due to the disruption of the β-galactosidase gene.

When selectable marker genes are used for the rapid analysis of the activity of promoters, they are called reporter genes. Green fluorescent protein (GFP) gene from the jellyfish, *Aequorea victoria*, is gaining popularity as a reporter gene. Because of chromophore formation and light emission by GFP, illumination of the site of expression of GFP with long-wave UV light or blue light leads to bright green fluorescence without any need for additional substrates (Cubitt et al., 1995; Yang et al., 1996).

Selectable Markers for Plants

A popular group of selectable markers used for selecting transformed plant tissue encode for herbicide resistance. In plant transformation the recombinant plasmids containing the plant selectable marker gene and the gene of interest are introduced into recipient plant cells by a variety of techniques (Potrykus, 1991). Genetically modified cells are then selected by growth on media containing inhibitory concentrations of the selecting agent; e.g., herbicide. Herbicide resistance marker genes can also be used under field conditions at the whole plant level to select transformed plants in segregating populations, to follow pollen dispersal of transgenic plant (Dale, 1992), outcrossing to wild species (Kerlan et al., 1992) and competitiveness and survival of transgenic plants in unmanaged habitats (Crawley et al., 1993).

The nonselective herbicide glyphosate which inhibits the essential enzyme 5-enolpyruvyl-3-phosphoshikimic acid synthase (EPSPS) in the chloroplast is a widely used selectable marker. An EPSPS gene from Agrobacterium sp. strain CP4 (Harrison et al., 1993) provides tolerance to glyphosate when its product is delivered to chlroplasts in various transgenic plants. Upon glyphosate treatment the transformed cells expressing the CP4 EPSPS are tolerant because the CP4 EPSPS is insensitive to glyphosate and thus continues to meet the plant's need for essential aromatic compounds in the presence of glyphosate.

Besides manipulation of the enzymes inhibited by glyphosate, the genes that code for the enzymes that inactivate the herbicide have been used as selectable markers. The enzyme glyphosate oxidase (GOX) catalyzes the cleavage of the C-N bond of glyphosate, yielding aminomethylphosphonic acid (AMPA) and glyoxylate as reaction products. The GOX utilizes oxygen as a substrate, does not produce hydrogen peroxide as a product of oxygen reaction and uses flavoprotein with FAD as a cofactor. The introduction into plants of glyphosate degradation enzyme (GOX) confers glyphosate tolerance to plants and augments the tolerance of transgenic plants already expressing a glyphosate tolerant EPSPS. Commercial cultivars of several crops tolerant to glyphosate have been produced by stably inserting both the CP4 EPSPS and GOX genes into their chromosomes.

The *bar* gene of *Streptomyces hygroscopicus*, which encodes the enzyme phosphinothricin acetyl transferase (PAT) conferring resistance to the herbicide, bialaphos or glufosinate, is also a useful marker for selection of transgenic plants (Vickers et al., 1996, Thompson et al., 1987; White et al., 1990).

Risk Assessment

The implications of the field release of transgenic plants have attracted global attention. With increasing international trade in agriculture, transgenic crops are under focus because the potential benefits must exceed the risk of introducing any hazards into the environment. In appraising the hazards of selectable markers, the following issues/questions are important:

- Toxicity or adverse effects of the selectable marker gene protein product or metabolites exposed to humans, wildlife, animals, beneficial insects, marine life, threatened and endangered species.
- In the case of antibiotic resistance selectable markers, is the gene likely to transfer from transgenic plants to pathogenic organisms? What is the clinical importance of the antibiotics that the selectable marker gene products inactivate? Are there alternative therapies and other antibiotics available besides the one inactivated by the selectable marker gene product to kill pathogenic bacteria with the selectable marker gene? Is the selectable marker gene expressed in bacteria or plants? If yes, what is the tissue specifiity and level of expression? How stable is the selectable marker gene product? Will the consumption of the transgenic plant product compromise therapeutic use of the selecting antibiotics? How widely spread is the selectable marker gene among current populations of human and animal pathogens and soil microbes? Is the gene already present in plasmids, integrons or transposons in microbes?
- Possibility of transfer of herbicide resistance selectable marker genes from transgenic crops into weedy relatives or other cultivated plants or cultivars and its impact on the use of herbicides, weed management, current agricultural practices, and the consequence of the use of the transgenic crop for the environment should also be assessed. The possibility of undesirable selectable marker gene flow from transgenic plant into wild species, changing relative position of the wild species in the ecosystem should be evaluated (Hokanson et al., 1997).
- Impact on present food web and ecosystem relations.

Since DNA from all living organisms are structurally similar, the presence of transferred DNA per se in food products poses no health risk. Similarly, large numbers of proteins of different structures and function are safely consumed and digested. Few proteins that are known toxins (e.g., cholera toxin) are not components of the human diet. The current range of marker genes used thus far do not code for any toxins and, therefore, are not considered dangerous. If the marker gene is derived from an allergenic source, then the allergenic potential of the expressed protein in the transgenic plant may be considered. If the marker gene protein inactivates an antibiotic used in human therapy, the potential for inactivation of an oral dose of the corresponding antibiotic exists. Inactivation may also require co-factors needed by the marker gene protein. It must also be resistant to digestion in the gastrointestinal tract or potentially consumed simultaneously with the antibiotic.

Herbicide tolerant transgenic crops are no more likely to become a weed than similar cultivars which could potentially be developed by traditional breeding techniques (Chopra et al., 1998; Paroda and Chadha, 1996). As long as the parent of the transgenic crop is not a serious, principal, or common weed pest, there is no reason to believe that a selectable marker would enable them to become weed pests. Although various definitions of the term "weed" have been proposed in scientific literature, the salient point is that a plant can be considered a weed when it is growing where humans do not want it to grow (Baker, 1965; de Wet and Harlan, 1975; Muenscher, 1980). Baker (1965) lists 12 common attributes that can be used to assess the likelihood that a plant species will behave as a weed. Keeler (1989) and Tiedje et al. (1989) have adapted and analyzed Baker's list to develop guidance to the weediness potential of transgenic plants. Both authors emphasize the importance of looking at the parent plant and the nature of the specific genetic changes.

Many cultivated plants that are being genetically engineered are not considered weeds and are unlikely to become a weed as a result of the introduction of a selectable marker. Many crops are highly inbred, well-characterized plants, and not persistent in undisturbed environments without human intervention. Although cultivated plant volunteers are common, they are easily controlled using many means; e.g., herbicides, biological and natural environmental factors, cultivation, and crop rotation practices. Many crops have only a few of the characteristics of successful weeds; e.g., production of abundant, long-lived seeds; vegetative propagation; and competition with other plant species in the undisturbed environment.

Genes that code for antibiotic inactivating enzymes have been widely used for cloning genes in *E. coli* and plants. The DNA sequences responsible for these markers get stably integrated in the plant genome and those DNA sequences, as such, have been a cause of concern for some individuals. Marketing of transgenic plants containing antibiotic resistance genes has been discussed (Flavell et al., 1992; Bryant and Leather, 1992; Gressel, 1992). Food and environmental safety of the kanamycin resistance gene and its gene product have been assessed (Nap et al., 1992). Each selectable marker gene should be assessed individually.

Selectable marker genes that are integrated in the genomes of transgenic plants have come under scrutiny because of their potential transfer to human pathogens and other organisms, both in the environment and in the intestine of humans and animals. The beginning of the polarization in scientific views is an ampicillin resistance (Amp) marker gene that was linked to an insecticide resistant trait and was integrated into the corn genome by Ciba Seeds. The interest in the horizontal transfer is heightened by the studies of Davies (1994) who announced long before recombinant DNA was the vogue that genes for antibiotic resistance were present in the antibiotic producing organisms. When polymerase chain reaction became available, Julian Davies used this sensitive technology to isolate genes for antibiotic inactivating enzymes from antibiotic preparations that are produced by fermentation of the producing organisms and used them as human therapeutics. These antibiotic resistance genes have been consumed with the antibiotic by the human patient and eventually may have been transferred to the human pathogens making them resistant to the antibiotic. The discovery of a rare enzyme, that inactivates chloramphenicol by phosphorylation in the producing organism and is not widely spread among clinical isolates, further complicates the issue. Chloramphenicol is produced by chemical synthesis for clinical use and is not contaminated by the DNA sequences that are responsible for phosphorylation of chloramphenicol. The chloramphenicol phosphorylating enzyme may not be widespread because chloramphenicol produced by fermentation contaminated with resistance genes was not used clinically.

Transgenic organisms are often no different from their parents in their ability to exchange genetic information with other organisms. Pollen from fertile, genetically modified plants may be transmitted via pollinators or wind to sexually compatible species. Subsequent fertilization could disseminate the marker gene into wild weedy relatives (Raybould and Gray, 1993). There is no selection pressure to select the transgenics containing antibiotic resistance genes, but this could be different for herbicide resistance if the selectable herbicide were a part of a weed control strategy for the weedy relative. Transfer distances of pollen from transgenic plants in a natural population should be the same as for wild type pollen. The production of hybrid seed between transgenics and wild relatives in the environment will also be affected by flowering times and sexual compatibility. The hybrid seeds may naturally be inviable, less fit, or infertile.

Glyphosate Resistance Gene: A Case Analysis

The herbicide, glyphosate, kills plant cells in the transformation process due to inhibition of the enzyme, EPSPS, (Steinrücken and Amrhein, 1980) which is found in microbes and plants but not in mammals (Levin and Sprinson, 1964; Cole, 1985). This also contributes to the selective action of glyphosate toward plants but not mammals.

The glyphosate resistance gene encodes an EPSPS enzyme that is insensitive to inhibition by glyphosate but has high affinity for phosphoenol pyruvate. This target enzyme is located in the chloroplast. It condenses phosphoenolpyruvic acid (PEP) and 3-phosphoshikimic acid (S3P) to 5-enolpyruvyl-3-phosphoshikimic acid, which is a precursor for the biosynthesis of aromatic amino acids, vitamins, and many secondary metabolites (Malik, 1989). As a consequence of the inhibition of aromatic amino acid biosynthesis, protein synthesis is disrupted resulting in the plant's death.

The expression of various mutant EPSPSs that were produced by site-directed mutagenesis in the plant failed to produce a commercially viable cultivar. This failure occured because all of the mutant EPSPS, in parallel to glyphosate tolerance, had decreased affinity for PEP. It was then reasoned that an organism that exists in nature and is tolerant of glyphosate may have evolved with an enzyme that has the desired characteristics; i.e., glyphosate tolerance, as well as high affinity for PEP. The naturally occurring soil bacterium, *Agrobacterium* sp. strain CP4, has the enzyme called CP4 EPSPS with high glyphosate tolerance and tight binding to PEP (Barry et al., 1992). The bacterial isolate, CP4, was identified by the American Type Culture Collection as an *Agrobacterium* species. There is no human or animal pathogenicity known from *Agrobacterium* species, nor is the EPSPS gene a determinant of *Agrobacterium* plant pathogenesis.

Agrobacteria occur almost worldwide in soils and in the rhizosphere of plants. *Agrobacterium* strains have also been reported in a number of human clinical specimens, but it is believed that these clinical *Agrobacterium* isolates occur either as incidental inhabitants in the patient or as contaminants introduced during sample manipulation (Kersters and De Ley, 1984).

Based on the kinetic parameters, and suitability for use in conferring glyphosate tolerance to crops, the gene for CP4 EPSPS was cloned from *Agrobacterium* sp. strain CP4 (Padgette et al., 1996). Based on the open reading frame in the nucleotide sequence, CP4 EPSPS is a 47.6-kDa protein consisting of a single polypeptide of 455 amino acids. Comparing the deduced amino acid sequences of CP4 EPSPS with the EPSPS from soybean, corn, petunia, *E. coli*, *Bacillus subtilis*, and *Saccharomyces cerevisiae* yields similarities of 51.2, 48.5, 50.1, 52.2, 59.3, and 53.5, respectively, and identities of 26.0, 24.1, 23.3, 26.0, 41.1,and 29.9%, respectively. The amino acid sequence homology between the CP4 EPSPS and EPSPSs typically present in plants and other bacteria (48.5 to 59.3% similar, 23.3 to 41.1% identical) is comparable to the homology between the EPSPSs from soybean and *B. subtilis* (55.6 similar, 30.1% identical). There is considerable divergence in the EPSPSs that are present in plants consumed by humans, and the divergence of the CP4 EPSPS sequence from typical food EPSPS sequence is of the same order. This can be taken to argue that CP4 EPSPS is not toxic to humans and other life forms. The CP4 EPSPS protein that is expressed in glyphosate-tolerant soybean, is rapidly digested and is not toxic to acutely gavaged mice (Harrison et al., 1996).

Amp Gene: A Case Analysis

The probability of horizontal transfer and the consequence of the ampicillin resistance gene (Amp) present in the genetically modified maize has been critically analyzed. The Amp gene encodes the ampicillin inactivating enzyme, β-lactamase. It is from the R6K plasmid (Datta and Richmond, 1966) and is part of a transposable element, TnA3, which is spread in a broad variety of plasmids/R factors (Heffron et al., 1975). The Amp gene was cloned to create the *E. coli* cloning vector plasmid, pBR322, (Bolivar et al., 1977) which is a progenitor of most popular cloning vectors today. The Amp gene can be inactivated by the insertion of cloned DNA or selected to maintain the inserted plasmid by growing the transformed bacterial populations in the presence of ampicillin. The Amp gene is now present in many plasmids derived from pBR322, including those of the pUC family which were used by CIBA Seeds in the development of insect-tolerant transgenic maize that is no longer a regulated article in the U.S.

Relationships Between the Various TEM Genes

TEM-lactamases are broad spectrum enzymes that inactivate most synthetic penicillins. A broad variety of TEM enzymes exist which are encoded by TEM genes that evolved from Amp. They differ in their specificity and kinetics constants. The catalytic activity of the TEM-lactamase is tolerant of amino-acid substitutions which is often responsible for modifications of their substrate specificity and electrophoretic behavior. Up to now, 28 different variants are known to be derived from the original TEM1 gene which is also called Amp (T. Hart: University of Liverpool; personal communication). Single point mutations are involved in evolution from TEM1 to TEM12 (Bradford, 1994). All genes that encode TEM enzymes and are related to Amp are relevant for the discussion of its use as a selectable marker.

Occurrence of TEM Genes on Transposons

So far, TEM-lactamase genes have been located on several hundreds of naturally existing different plasmids (Sykes and Mathew, 1976). Most of these plasmids that carry TEM1 are conjugative, and they transfer in soil from one organism to another as a result of their very active transposition by the TnA family of transposons (Heffron et al., 1975). The presence of the gene on transposons (eg., Tn1, Tn2, Tn3, Tn1701, Tn901 etc.) and integrons is described in a number of publications (Yamamoto et al., 1982, Bunny et al., 1995). Consequently, these genes are spread in numerous gram negative bacterial species belonging to the genera *Hemophilus*, *Escherichia*, *Neisseria*, *Citrobacter*, *Salmonella*, *Proteus*, *Klebsiella*, and *Pseudomonas* which exchange plasmids by conjugation.

The TEM1 gene has been introduced into *Bacillus subtilis* (Ulmanen et al., 1985) and *Anacistis nidulans* (van der Hondel et al., 1980). This gene does not occur naturally in these two organisms. The real test of the horizontal transfer of the ampicillin gene from corn to soil microbes could be if ampilicillin-harboring *Bacillus subtilis* and *Anacistis nidulans* could be isolated from the transgenic cornfields.

Incidence of the TEM-lactamase Gene in Human and Animal Pathogens

A survey of more than 300 references in Medline and VETU sources indicates the following:

- TEM1 is present in numerous bacterial species isolated from diseased patients, including the genus *Hemophilus*, *Escherichia*, *Neisseria*, *Citrobacter*, *Salmonella*, *Proteus*, *Klebsiella*, and *Pseudomonas* (Sykes and Mathew, 1976). This resistance is found in all the five continents (Neu, 1992).

- TEM1 accounts for approximately 50% of the cases of resistance to penicillins in gram negative bacteria associated with human pathology, all species confounded (Neu, 1992) and irrespectively of their geographical origins, with an increased prevalence in hospitals.

- Though antibiotic resistance is monitored quite extensively in animals, the nature of the genes involved is not as precisely investigated as it is in human medicine. However, the importance of TEM-lactamase in animal epidemiology may be inferred from what is known in human epidemiology.

- Importance of the flux of pathogenic microbes between animals and humans has been argued. However, such a flux is supported in the case of *Salmonella* infections, indicating that R-factors can be exchanged between animal and human micro flora (Walton, 1983, 1988; Linton, 1984, 1983).

Horizontal Gene Transfer: Transgenic Maize to Rumen Microbes

There is no scientific evidence for the occurence of direct gene transfer of DNA present in the gastrointestinal tract of animals to its microflora. The transgenic maize will be used as cattle feed and possibility of the subsequent transfer of an Amp gene from transgenic corn to the animal intestinal microflora has been raised. The concern is that the Amp gene will move from the plant tissue to microorganisms that commonly inhabit the animal gut, consequently reducing the efficacy of ampicillin-antibiotic therapy in that animal. However, for such a horizontal Amp gene transfer numerous natural barriers exist. Even if the transfer of a functional Amp gene occured, widely used broad spectrum β-lactams and other antibiotics that are not inactivated by the enzyme encoded by the Amp gene can be used to kill ampicillin resistant infections.

The successful transfer of plant DNA into microorganisms is directly impacted by the integrity of the plant DNA. A DNA fragment with the intact Amp gene and controlling elements must be transferred and integrated into bacterial chromosome to be expressed. Regardless of whether the harvested tissue is ensiled or fed as green chop to cattle, the half-life of plant genomic DNA in either case will be extremely short. As most maize green tissue will be ensiled prior to feeding, the low pH and degradative enzymes present in silos will very likely result in rapid

DNA degradation. DNA not degraded to oligonucleotides prior to consumption will be subject to the harsh environment of the rumen and gut.

DNA degradation by plant nucleases. Plant cells have an abundance of highly active nucleases that will serve to severely compromise the integrity of plant DNA upon cell lysis (Ausubel, 1987). Following disruption of plant cells, it is necessary to suspend the tissue immediately in strong denaturing agents in order to isolate plant DNA with minimal nucleolytic digestion (Ausubel, 1987; Wang et al., 1996). Integrity of plant DNA in the absence of protein denaturing agents is hard to maintain. Such stabilizing agents are absent in the stomach of an animal. During DNA isolation, within one hour of tissue disruption without protein denaturants and nuclease inhibitors, all DNA is degraded to fragments of less than 500 base pairs, while the majority was less than 150 base pairs. The smallest contiguous DNA fragment from pUC18 (a commonly used research plasmid) that could contain the Amp gene is ~900 base pairs, while the fragment size necessary to contain both the origin of replication (ori) and the Amp gene is approximately 1600 base pairs. The experimental data indicate that plant-derived nuclease activity alone will be adequate to eliminate DNA fragments of 1600 base pairs from the plant tissue.

DNA degradation by microbial nucleases. Any rare DNA fragments that might escape degradation during the ensiling process or by plant nucleases would be subject to digestion by extracellular nucleases from ruminal and gut bacteria. Ruminal fluid and the small intestine contain bacterially secreted or lytically released nucleases that rapidly and completely degrade single-stranded and double-stranded DNA (Flint and Thomson, 1990; McAllan, 1980). Utilizing bacteria-free ruminal fluids, Russell and Wilson (1988) demonstrated substantial DNA degradation immediately upon assay (0 time) at a temperature of 0 degree C. These authors claim that due to the high level of ruminal microbial nucleases, it is highly unlikely that transformation would occur in the rumen under natural conditions, a view shared by Flint (1994). Similar results were obtained using pure cultures of ruminal bacteria (Flint and Thomson, 1990). Morrison (1996) estimates the half-life of DNA in ruminal fluid to be from 1 to 13 minutes. Plant DNA surviving the upper chambers of the ruminant stomach would be exposed to the acidic pH of the abomasum. Under low pH conditions, the DNA undergoes rapid depurination and strand fragmentation (Maniatis et al., 1982). Ruminants are unique among mammals in the ability to digest large amounts of nucleic acids in their diet due to microbial nuclease activity in the rumen stomach and lower gut.

Despite the evidence that DNA released from the plant nucleus will be extensively degraded by nucleases, it is recognized that DNA can be partially protected from nucleolytic digestion when associated with particulate (e.g., soil, plant) matter (Lorenz and Wachernagel, 1994). While the DNA fragment would be protected from further degradation, the high affinity of DNA to organic substrates and minerals also renders it unavailable for binding to proteins on the bacterial cell surface, a prerequisite for DNA uptake into the bacterial cell (Stewart and Carlson, 1986).

In addition to the barriers of nuclease digestion and cell uptake, the DNA would be exposed to intracellular restriction endonucleases which are common in ruminal bacteria (Morrison, 1996). The action of these enzymes would further serve to compromise the structural integrity and maintenance of the DNA in the bacterium.

The process of DNA uptake by natural transformation. Another component necessary for the successful transfer of an intact ori-Amp fragment is that the naked DNA must gain entry into the bacterium by a process known as natural transformation. This process requires recipient cells to be in a physiologically competent state prior to transformation (Stewart and Carlson, 1986). There is no evidence that DNA transfer in the rumen occurs by naked DNA transformation or any other mechanism of gene transfer, including conjugation (Morrison, 1996).

Researchers have examined the significance of competency to transformation. While transformation occurs with high frequency in an *E. coli* strain that is artificially induced to be competent, there were no transformants from the noncompetent cell line, even though a covalently closed circular form of the plasmid (puC19) was utilized. The competency required for natural transformation is a highly complex process, markedly affected by growth conditions, age of cells, and environmental conditions (Stewart and Carlson, 1986). There is no published scientific evidence that ruminal or gut bacteria are capable of a natural competent state or scientific evidence of exogenous DNA transfer into ruminal or gut bacteria in their respective environments.

The integration of plasmid DNA into the chromosomes of maize occurs via a random recombination event. This precludes the looping out and recircularization of the plasmid DNA from the plant chromosome. Thus, there would be no opportunity for an intact plant derived plasmid molecule to present itself to a bacterium for uptake. A highly improbable event with limited nuclease digestion could generate a fragment of DNA that harboured both the Amp gene and origin of replication. However, this not necessarily would yield a sucessful transformation of rumen bacteria.

All current theoretical models and scientific evidence suggest that bacterial cells cannot be transformed by monomeric forms of plasmids (Stewart and Carlson, 1986). Rather, multimeric plasmid forms (double-stranded dimers and higher oligomers) aggregate at binding sites on the bacterial cell surface. This is followed by complete nucleolytic digestion of one strand of each plasmid molecule and partial digestion of the second strand. Upon entry into the cell, the overlapping (homologous) sequences of the single-stranded fragments hybridize and recircularize, prior to strand synthesis (Lorenz and Wackernagel, 1994; Stewart and Carlson, 1986). A very important component in the uptake process is the presence of multimeric forms of homologous DNA (plasmid) sequences at the same binding site on the cell surface. Therefore, in order for there to be bacterial uptake of an intact ori-Amp fragment, multiple copies of this fragment would have to emanate from the plant genome and aggregate at a binding site, a very improbable scenario.

Uptake of DNA by bacteria has been shown to occur without regard to specific nucleotide sequences (Hanahan, 1983). Each linear fragment of plant DNA presented to a rumen or gut bacterium has similar access to binding sites on the cell surface and similar opportunities for uptake. Norgard et al. (1978) and Hanahan (1983) demonstrated that fragmented DNA molecules compete for binding sites and were able to significantly reduce *E. coli* transformation by pBR 322 by adding increasing amounts of linear (heterologous) DNA. A 50,000-fold mass ratio of linear DNA to plasmid reduced transformation efficiency 10-fold. In a transgenic maize plant with a single insert (one locus), there is estimated to be a 600,000-fold mass ratio of heterologous linear DNA to integrated plasmid in maize plant cells (assuming the plant DNA fragments to 10 kb pieces). This overwhelming amount of competitive DNA, together with the stringent requirements necessary to achieve natural transformation, would serve to functionally eliminate the opportunity for a successful transformation event.

Scientists have experimentally evaluated the potential of a chromosomally integrated plasmid sequence (containing an Amp gene) to transform a bacterial cell. Transformation experiments were conducted utilizing both nuclease-degraded and intact plant DNA from a genetically modified maize plant. DNA was incubated with a highly competent strain of *E. coli* to maximize the opportunity to produce a transformant. Upon completing the transformation protocol, the bacteria were plated onto ampicillin-containing medium. No ampicillin-resistant colonies arose, indicating that even under ideal experimental conditions, a successful transformation does not occur above a frequency of 1 in 6.8 x 10(19). Thus, in the event that an ori-Amp fragment survives the hostile ensiling and/or digestive processes, there is no scientific evidence that supports the possibility of successful uptake of intact, functional plant genes into a ruminal bacterium. In the highly unlikely event that plant DNA avoided nuclease digestion and gained entry into a cell, the DNA would have two possible fates—

- autonomous replication (requiring an origin of replication)
- stable integration into the bacterial chromosome

Autonomous replication of a transforming plasmid requires that the host DNA replication enzymes and associated cofactors recognize the origin of replication present in the plasmid. The origin of replication in pUC (as well as pBR322, and ColEl) is only recognized by replication enzymes present in a small group of bacteria in the Family Enterobacteriaceae (Kues and Stahl, 1989). Plasmids derived from pUC do not replicate in bacteria that are phylogenetically unrelated to the Enterobacteriaceae group of gram-negative bacteria. Plasmids derived from pUC are incapable of replicating in gram-positive bacteria and do not replicate in any ruminal or gut bacteria except for transient strains of *E. coli* and *S. typhimurium*. Further, the possibility of autonomous replication is moot since DNA taken up by natural transformation, assuming it survives restriction enzymes (which are abundant in most natural isolates), will be in the form of linear molecules and will have to integrate by homologous recombination.

Stable integration of foreign DNA sequences in bacteria occurs nearly exclusively by homologous recombination (Wilson, 1985). Integration requires significant DNA homology between the transforming DNA molecule and sequences in the chromosome. The likelihood of there being sufficient homology between plasmid sequences and the chromosomal DNA of ruminal bacteria is minuscule. The numerically predominant bacteria in the rumen are either gram-positive or members of the Prevotella-Fibrobacter groups of gram-negative bacteria. Those latter gram-negative bacteria are even less genetically related to the

E. coli-Pseudomonas group than the gram positive. Finally, in order to achieve homologous recombination between ori-Amp from pUC and the recipient chromosome, there must already be an ori-Amp gene present in the chromosome. The probability for a natural transfer of the specific ori-Amp sequence from a genetically modified maize to a rumen or gut pathogen with prerequisite integration, maintenance, and expression, collectively necessary to compromise the effectiveness of an otherwise effective β-lactam antibiotic, is, at best, so remote that its occurence is irrelevant to an assessment of food or feed safety, and more likely, impossible.

Clinical consequence of Amp gene transfer. The rumen is inhabited by a diverse group of microorganisms. These cells compete for nutrients, and the expression of an unneeded gene such as the Amp gene could result in the bacterium being less competitive and unable to attain high numbers in the rumen. In fact, an organism expending energy by expressing an Amp gene would most likely be at a competitive disadvangage in the rumen environment unless the animal was treated with ampicillin or similar antibiotics.

While β-lactum antibiotics are very widely used, it is important to note that there are several subclassifications of β-lactam antibiotics, and they differ markedly in their sensitivity to lactamases (The Merck Veterinary Manual, 1991). For example, while some narrow spectrum β-lactams (e.g., benzylpenicillin and penicillin V) are sensitive to β-lactamases, oxacillin, cloxacillin, dicloxacillin, flucloxacillin, temocillin, methicillin, and nafcillin (all β-lactams) are insensitive to β- lactamases. The widely used extended-spectrum β-lactams, while inactivated by many β-lactamases already present in the environment, are not inactivated by the β-lactamase encoded by the Amp gene present in pUC, nor is the pUC β-lactamase capable of inactivating the more commonly used second and third generation cephalosporin class of antibiotics. Further, there are several compounds (e.g., clavulanic acid) that are competitive inhibitors of β-lactamase enzymes. When used in combination with penicillins, such antibiotics are protected from enzymatic hydrolysis and are fully active towards previously resistant bacteria. Based upon these considerations, the clinical consequence of Amp gene expression in ruminal bacteria is noneventful.

Conclusion. The Amp gene in genetically modified maize is not of significant risk because—

- An intact ori-Amp sequence is extremely unlikely to survive the hostile conditions during the ensiling process.
- Even if such an event were to occur, an intact ori-Amp sequence is extremely unlikely to survive intact the hostile environment of an animal gut or rumen.
- Even if this would happen, it is highly improbable that such DNA would find its way to competent bacterial cells and undergo successful natural transformation.
- If such DNA were nevertheless to survive intact and be successfully transformed, it would be virtually impossible to autonomously replicate or be maintained and expressed in a ruminal or gut organism.
- Even if the above four steps were to take place against such overwhelming odds, there would be no clinical impact.

Horizontal Gene Transfer

Advances in recombinant DNA technology allow transfer and expression of any gene from anywhere in the living kingdom into plants. This has extended the gene-pool for crop improvement well beyond the boundaries defined by sexual incompatibility. There is public scrutiny of the environmental and evolutionary consequences of the release of genetically engineered organisms into the environment and its impact on foods. The possibility of free dispersal of novel genes from the confines of a genetically engineered organism to taxonomically unrelated life forms with consequences for ecosystems poses parallel and intriguing questions. Movement of genetic information from plants to microorganisms and vice versa via nonsexual processes is called horizontal gene transfer. Horizontal gene transfer is central to all environmental impact assessments prepared by regulatory agencies that relates to genetic stability and is of foremost importance to biosafety. There is currently considerable skepticism about the occurrence of horizontal gene transfer and to its importance. Naturally occurring horizontal gene transfer is difficult to prove. Doolittle et al. (1990) has used sequence data from a wide variety of organisms to argue that *E. Coli* acquired a second glyceraldehyde-3-phosphate dehydrogenase gene from some eukaryotic host.

Horizontal gene transfer between remotely related species in nature has not been aggressively explored. Evolutionary biologists have used phylogenetic intuitions to deduce and speculate about movement of genetic material in the living kingdom. Hard facts are practically nonexistant. Chromosomes are committed to sexual vertical gene transfer while plasmids move both vertically and horizontally. Even though epidemiological surveys of drug resistant genes in human pathogens and phylogenetic analysis of extant organisms show the role of horizontal gene transfer in their history, there is no solid experimental evidence showing independant movement of chromosomal genes from eucaryotes to procaryotes. However transfer of plasmids from *E. Coli* to *Streptomyces*, yeast, and mammalian cells is well established. Transfer of genetic information from *Agrobacterium* to plants involving tumor inducing plasmid is the best studied example of horizontal gene transfer from a bacterium to the plant (Zymbriski).

The residues of transgenic crops are a potential source for transgenic DNA which can be released into the soil and taken up by soil microbes (BilVoet and Nap, 1992). The frequency of gene transfer from transgenic crop to a soil microbe estimated at 5.4×10^{-4} is similar to the rate of natural mutations under selection pressure and additional characterization of the transformed sequence in the recipient is needed to make this study convincing. Hoffman et al. (1994) have documented evidence for transfer of the hygromycin resistance gene from the transgenic plant of Datura innoxia via soil to *Aspergillus niger*. There are not many other claims of horizontal gene transfer from eucaryotes to procaryotes (Prins and Zadoks, 1994, Evenhuis and Zadoks, 1991).

Soil microbes and antibiotic producing industrial strains already contain many types of resistance genes that encode antibiotic inactivating enzymes (Davies 1994). Antibiotic preparations could contain the resistnce genes as a minor contaminant (Webb and Davies, 1994). As a matter of fact, most selectable marker genes, especially for resistance to antibiotics and herbicides, are so widely spread among soil microbes that their transfer from transgenic plants to soil microbes should not create novel gene combinations that already do not exist. Resistance to chloramphenicol, streptomycin, ampicillin, and neomycin is located on transmissible plasmids and already widespread among bacteria that cause disease in humans. The consumption of transgenic crops with DNA sequences encoding for antibiotic resistance should not result in the evolution of any novel pathogens drastically different from those that already exist. A high degree of similarity between the polymerase domain of bacterial DNA polymerase I and two distinct human sequences putatively involved in DNA repair suggests recruiment of bacterial genes by mammalian cells (Sonnhammer et al., 1997).

Sexual Transfer of Selectable Marker Genes

Even though potential of transfer of a marker gene to a weedy relative exists, it will depend on the degree of sexual compatibility between donor and recipient species as well as physical distance between the two (Table 3). The donor and recipient plants should produce receptive flowers at the same time in the presence of any necessary pollinating agent. Detemination of the likelihood and possible consequences of transfer of introduced genes (transgene) by sexual hybridization to plants of the same or related species is part of the risk assessment conducted before transgenic plants are released into the environment.

The extent of movement of transgenes by pollen to adjacent potato plants was found to be very limited. Transgenes did not move from the potatoes to *Solanum dulcamara* and *Solanum nigrum* (Mcpartlan and Dale, 1994). Plants resistant to antibiotics will not become weeds since there is no selection for antibiotic resistant weeds in nature and antibiotic resistance does not provide any competitive advantage limiting its impact on weediness. However, herbicide resistance marker genes with respect to weediness needs evaluation. The resistance to widely used herbicides will gradually evolve among weeds and plants, even in the absence of transgenic crops with herbicide tolerance genes. Repeated use of herbicides is eventually to the select weeds that are resistant to herbicides. The acquisition of a single selectable herbicide-tolerant marker gene is unlikely to convert a cultivar into a weed. Unless the crop is a frequent volunteer in a field where the herbicide is often used to control weeds, a plant must acquire a number of characteristics before it becomes a weed in the absence of herbicide selection (Keeler, 1989).

Long-term ecological consequences of releasing genetically modified plants containing herbicide-tolerance marker genes are a concern (Crawley et al., 1993; Duke, 1996). Wheat is tolerant of dichlorsulfuron and is often grown in rotation with oilseed rape. Introduction of chlorsulfuron resistance into rape could result in volunteer oilseed rape becoming a weed problem in wheat crops (Gressel, 1992b) when dichlorsulfuron is used as a herbicide. Transgenic oats with glufosinate resistance (bar) could be volunteers in wheat fields (Gressel, 1992) but they can be controlled with other herbicides. Glyphosate can eliminate volunteer crops such as potato, beet, and cereals. Tolerance to glyphosate in volunteer crops will require alternative herbicides for volunteer control. If herbicide-tolerance is engineered into Sorghum, it could move into Johnson grass and could, therefore, have ecolgical consequences.

Alternatives to Selectable Markers

The use of antibiotic resistance markers is a convenience but not an absolute requirement for cloning genes in *E. coli*. The gene sequences for the biosynthesis of amino acids, purines, and pyrimidines are routinely used for cloning genes in yeast and fungi (Botstein and Fink, 1988). Such markers are especially appealing because the construction of the recipient host by homologous recombination with deletion of selectable marker sequences is now possible. Similar selectable markers can be used for cloning genes in *E. coli* thus bypassing the use of the controversial antibiotic resistance. The selectable marker gene for antibiotic resistance can also be cleaved out of the recombinant plasmid before it is introduced into the plant tissue. Cotransformation with a plasmid that can be used to select the transformed cells and later eliminated is also a possibility (Vickers et al., 1996).

For plants, the use of herbicide marker genes may be preferred over antibiotic resistance genes (Wehrmann et al., 1996). Existing marker genes from modified organisms may be eliminated by the Cre/lox recombination system (Dale and Ow, 1991). Many plants can now be regenerated from single cells. The direct injection of DNA into single isolated plant cells followed by regeneration into a transgenic plant is a cumbersome alternative, even when it is possible.

U.S. Regulatory Perspective on Selectable Markers

In response to a petition by Monsanto Co., the U.S. Environmental Protection Agency (EPA) established an exemption from the requirement of a tolerance for residues of neomycin phosphotransferase II (NPTII) and the genetic material for its production (Federal Register 59: 187 September 28, 1994 p 49351). Under section 408 of the Federal Food Drug and Cosmetic Act (21 U.S.C. 346a) as amended in 40 CFR part 180, Ciba-Geigy Corp. received from EPA an exemption from the requirement of a tolerance for the phosphinothricin acetyl transferase as produced in corn by the bar gene and its controlling sequences as found on plasmid vectors pCIB 3064. Similar exemption was also awarded to Northrup King Company for PAT and the genetic material for its production

(Plasmid vector pZO1502; EPA August 2, Federal Register, final 40 CFR 180.1175 ; Federal Register 61; 31 Feb. 14, 1996 p. 5772). Furthermore, EPA has issued a final rule establishing an exemption from the requirement of a tolerance for residues of Enolpyruvylshikimate phosphate synthase in all plants (EPA August 2, Federal Register, Final 40CFR180.1174).

Because of the ambiguous nature of the controlling statutory provisions and unique nature of plant-pesticides and substances as selectable markers, EPA concludes that a substance to confirm the presence of the genetic material necessary to produce the active ingredient are not components of a pesticide. Should EPA decide that substances such as selectable markers are not inert ingredients or pesticide components, FDA, rather than EPA, would have direct jurisdiction over the presence of those substances in food products. Should EPA decide that substances, and related genetic material, used to confirm and ensure the presence of the plant pesticide should not be classified as part of a pesticide, the regulatory text in the final rule under FIFRA and FFDCA would be modified to reflect this decision, including defining the plant-pesticide product as the plant-pesticide active ingredient (Federal Register, 61:141, Monday, July 2, 1996).✧

Selected Reading

Ausubel, F. (ed.), 1987 Current protocols in molecular biology. Wiley & Sons.

Baker, H. G.and Stebbins, G. L., (eds.), 1965. In: The genetics of colonizing species, p. 147–168. Academic Press, New York and London.

Bolivar F. et al., 1977. Gene 2:95–113.

Botstein, D. and Fink, G. R., 1988. Science 240:1439-1443.

Barry, G., Kishore, G., Padgette, S., Taylor, M., Kolacz, K., Weldon, M., Re, D., Eichholtz, D., Fincher, K., and Hallas, L., 1992. In: B.K. Singh, H.E. Flores, and J.C. Shannon (eds.). Biosynthesis and molecular regulation of amino acids in plants, p. 139-145. American Society of Plant Physiologists, Madison, Wisconsin.

Barry, G.F., Taylor, M.L., Padgette, S.R., Kolacz, K.H., Hallas, L.E., della-Cioppa, G., and Kishore, G.M., 1994. Cloning and expression in Escherichia coli of the glyphosate-to-aaaminomethylphosphonic acid degrading activity from Achromobacter sp. strain LBAA. Monsanto Technical Report MSL–13245, St. Louis.

Bradford P.A., 1994. Antimicrobial Agents and Chemotherapy 38:761–766.

Bunny, K.L., Hall, R.M. and Stokes, H.W., 1995. Antimicrobial Agents and Chemotherapy 39:686–693.

Chopra, V.L., Singh, R.B., and Varma, A. (eds.), 1998. Crop productivity and sustainability shaping the future, 1111 pp. Proceedings of the 2nd International Crop Science Congress. Oxford & IBH Publishing, New Delhi.

Cole, D.J., 1985. Mode of action of glyphosate—a literature analysis, p. 48–74. In: Grossbard, E. and Atkinson, D. (eds.). The herbicide glyphosate. Buttersworths, Boston.

Crawley, M J., Hails, R.S., Rees, M., Kohn, D., and Buxton, J., 1993. Nature 363:620–623, London.

Crockett, L., 1977. Wildly successful plants: North American weeds, 609 pp. University of Hawaii Press, Honolulu, Hawaii.

Cubitt, A.B., Heim, R., Adams, S. R., Boyd, A.E., Gross, L.A., and Tsien, R.Y., 1995. Understanding, improving, and using green fluorescent proteins. Trends in biochemical sciences, p.448-456. V. 20, No. 11.

Dale, P.J., 1992. Plant Physiol 100:13-15.

Dale, E.C; Ow, D.W., 1991. Proc. National Acad. Sci. U.S. 88, 10558–10562.

Datta, N. and Richmond, M.H., 1966. Biochemical Journal 98:204–209.

Davies, J., 1994. Science 264:375–380.

de Wet, J.M.J.and Harlan, J.R. 1975. Economic Botany 29:99107.

Doolittle, R.F., Feng,D.F., Anderson, K.L., and Alberro, M.R., 1990. J. Molecular Evolution 31:383-388.

Duke, S.O., 1996. Herbicide resistant crops. CRC Press, New York.

Evenhuis, A. and Zadoks, J.C., 1991. Euphytica 55:81–84.

Flavell R., Dart, E., Fuchs, R.L, and Fraley, R.T., 1992. Bio/Technology 10:141–144.

Flint, H.J., Thomson, A.M., 1990. Letters in Applied Microbiology 11:18–21.

Flint, H.J., 1994 FEMS Microbiology Letters 121:259–268.

Galinat, W. C., 1988. In: Sprague, G. F., Dudley, J. W., (eds). Corn and corn improvement, Third Edition, p. 1–31. American Society of Agronomy, Crop Science Society of America, and Soil Science Society of America. Madison, WI.

Gould, F., Follett, P., Nault, B., Kennedy, G.G., 1994. In: G.W. Zehnder, M.L. Powelson, R.K. Jansson, and K.V. Raman eds.). Advances in potato pest management: biology and management, p. 255-277. American Phytopathological Society, St. Paul, MN.

Gould, F.W., 1968. Grass systematics, 382 pp. McGraw Hill, New York et alibi.

Gressel, J., 1992. Trends in Biotechnology 10:382.

Hallauer, A. R., Russell, W. A., Lamkey, K. R., 1988. In: G.F. Sprague and J.W. Dudley (eds.), Corn and corn improvement, Third Edition, p. 463-564. American Society of Agronomy, Crop Science Society of America, and Soil Science Society of America. Madison, Wisconsin.

Hanahan, D., 1983. J. Mol. Biol. 166:557–580.

Harding, K., 1995 Biosafety of selectable marker genes. CAB International 47N.

Harrison, L. A., Bailey, M. R., Naylor, M. W., Ream, J. E., Hammond, B. G., Nida, D. L., Burnette, B. L., Nickson, T. E., and Mitsky, T. A., 1996. Recombinant 5-enolpyruvylshikimate-3-phosphate synthase in glyphosate-tolerant soybeans is digestible. J. Nutr. 126(3):728–740.

Haughn, G.W., Smith, I., Mazur, B., and Somerville, C., 1988. Molecular and Gen Genetics 2ll:266–271.

Heffron F., 1975. J. Bacteriology 122:250–256.

Helmer, G., Casadaban, M., Bevan, M., Kayes, L., and Chilton, M.D., 1984. Bio/Technology 2:520–527.

Herrera-Estrella, L., Blodc, M. de, Messesvs, E., Hemalsteens, J.P., Montagu, M. van, and Schell, J., 1983a. EMBO Journal 2:987–995.

Herrera-Estrella, L., Depicker, A., Montagu, M. van, and Schell, L., 1983b. Nature 303:209-213.

Hitchcock, A. S., Chase, A., 1951. Manual of the Grasses of the United States, 1051 pp. U.S. Government Printing Office, Washington, DC.

Hoffman, T., Golz, C., and Schieder, O., 1994. Current Genetics 27:70–76.

Hokanson, S.C., Grumet, R., and Hancock, J. F., 1997. Effect of border rows and trap/donor ratios on pollen-mediated gene movement. Ecol. Applications 7(3):1075–1081.

Holm, L., Pancho, J.V., Herbarger, J.P., and Plucknett, D.L., 1979. A Geographical Atlas of World Weeds, 391 pp. John Wiley and Sons, New York.

Keeler, K., 1989. Bio/Technology 7:1134–1139.

Kerlan, M.C., Chevre, A.M., Eber, F., Baranger, A., and Renard, M., 1992. Euphytica 62:145-153.

Kersters, K., and De Ley, J., 1984. In: N.R. Kreig and J.G. Holt (eds.), Bergey's Journal of Systematic Bacteriology 1:244-248. Williams and Wilkins, Baltimore.

Kues, U., Stahl, U., 1989. Microbiological Reviews 53:491–516.

Levin, J.G. and D.B. Sprinson, 1964. The enzymatic formation and isolation of 5-enolpyruvylshikimate-3- phosphate. J. Biol. Chem. 239: 1142–1150.

Linton A.H., 1983. Bull. WHO 61:383–394.

Linton A.H., 1984. British Medical Bulletin 40:91–95.

Lorenz, M.G. and Wackernagel, W., 1994. Microbiological Reviews 58:563–602.

Malik, V.S., 1979. Genetics of applied microbiology. Advances in Genetics 20:37–126.

Malik, V.S., 1986. Genetics of secondary metabolism in biotechnology, Volume 4, p. 39–68. H.J. Rehm and G. Reed (eds.), Springer-Verlag, New York, VCH Verlagsgesellschaft, Weinheim.

Malik, V.S. and Wahab, S.Z., 1993. Versatile vectors for expressing genes in plants. J Plant Biochemistry and Biotechnology 2:69–70.

Maniatis, T., Fritsch, E.F., and Sambrook, J., 1982. Molecular cloning. CSH Press.

McAllan, A.B., 1980. British J. of Nutrition 44:99–112.

McPartlan, H.C., Dale, P.J., 1994. Transgenic Research 3:216–225.

Morrison, M., 1996. Do ruminal bacteria exchange genetic material? (In press).

Mosher, R.H., Camp, D.J., Yang, K., Brown, M.P., Shaw, W.P. and Vining, L.C., 1995. Inactivation of chloramphenicol by o-phosphorylation. J.Biol Chem. 270:27000–27006.

Muenscher, W. C., 1980. Weeds, 586 pp. Second Edition. Cornell University Press, Ithaca and London.

Nakamura, L.K., 1994. International Journal of Systematic Bacteriology 44:125–129.

Nap, J.P., Bijvoet, J., Stiekema, W.J., 1992. Transgenic research I: 239–249.

Neu, H.C., 1992. Science 257:1064–1073.

Norgard, M.V., Keem, K., and Monahan, J., 1978. Gene 3:279–292.

Padgette, S.R., Kolacz, K.H., Delannay, X., Re, D.B., La Vallee, B.J., Tinius, C.N., Rhodes, W.K., Otero, Y.I., Barry, G.F., Eichholtz, D.A., Peschke, V.M., Nida, D.L., Taylor, N.B., and Kishore, G. M., 1995. Development, identification, and characterization of a glyphosate-tolerant soybean line. Crop Science 35(5):1451–1461.

Padgette, S.R., Re, D.B., Barry, G.F., Eichholtz, D.E., Delannay, X., Fuchs, R.L., Kishore, G.M., and Fraley, R.T., 1996. New weed control opportunities: development of soybeans with a Roundup ready gene, p. 53-84. In: S.O. Duke (ed.), Herbicide-resistant crops: agricultural, environmental, economic, regulatory, and technical aspects. CRC Press, Inc., Boca Raton, Florida, USA; London, England.

Paroda, R.S. and Chadha, K.L. (eds.), 1996. Fifty years of crop science research in India, 796 pp. Indian Council of Agricultural Research, New Delhi, India.

Potrykus, I., 1991. Annual Review of Plant Physiology and Plant Molecular Biology 42, 205, 225.

Prins, T.W. and Zadoks, J.C., 1994. Euphytica 76:133–138.

Raybould, A.E., Gray, A.J., 1993. Journal of Applied Ecology 30, 199–219.

Russel, J.B., Wilson, D.B., 1988. J. Nutrition 118:271–279.

Schubert, R., 1994. Mol. Gen. Genet. 242:495–504.

Smulevitch, S.V., Osterman, A.L., Shevelev, A.B., Kaluger, S.V., Karasi, A.I., Kadryov, R.M., Zagnitko, O.P., Chestukhina, G.C., and Stepanov, V.M., 1991. FEBS Letters 293:25–28.

Sonnhammer, E.L. and Wooten, J.C. (1997). Widespread eukaryotic sequences highly similar to DNA polymerase I looking for functions. Curr. Biol 7: R463–R465.

Steinrucken, H.C. and Amrhein, N., 1980. The herbicide glyphosate is a potent inhibitor of 5-enolpyruvyl shikimic acid-3phosphate synthase. Biochem. Biophys. Res. Commun. 94:1207–1212.

Stewart, G.J., Carlson, C.A., 1986. Ann. Rev. Microbiol. 40:211–235.

Sykes, R.B. and Mathew M.J., 1976. Antimicrob. Chemoth. 2:115–157.

Sykes, R.B. and Mathew M.J., 1976. Antimicrob. Chemoth. 2:115–157.

Tabashnik, B.E., 1994b. Annual Review of Entomology 39:47–79.

Tabashnik, B.E., 1994a. Proceedings of the Royal Society of London, Series B Biological Sciences 255, 1342:7–12.

Thompson, A.R., Gore, F.L., 1972. Journal of Economic Entomology 65:1255–1260.

Thompson, C.J., Movva, N.R., Tizard, R., Crameri, R., Davies, J.E., Lauwereys, M., Botterman, J., 1987. Characterization of the herbicide-resistance gene bar from streptomyces hygroscopicus. EMBO Journal 6:2519–2523.

Tiedje, J.M., Colwell, R.K., Grossman, Y.L., Hodson, R.E., Lenski, R.E., Mack, R.N., Regal, P.J, 1989. Ecology 70:298–315.

Ulmanen, I., 1985. J. Bacteriol. 162:176–182.

Van der Hondel, C., 1980. Proc. Natl. Acad. Sci. USA 77:1570–1574.

Vickers, J.E., Graham, G.C., Henry, R.J., 1996. Plant mol. biol. reporter 14:363-368.

Walton, J.R., 1988. The Veterinary Record 122:249–251.

Walton J.R., Zbl. Vet. Med. A.(1983 30:81–92.

Wang, X., Wang, Z., and Zou, Y., 1996. Plant Mol. Biol. Reporter 14:369-373.

Webb, V. and Davies, J.T., 1994. Tibtech 12:74–78.

Wehrmann, A., Vliet, A.V., Opsomer, C., Botterman, J., Schulz, A., 1996. Nature Biotechnology 14:1274–1278.

Wilson, J.H. (ed.), 1985. Genetic Recombination. Benjamin/Cummings, Publishers.

Yamamoto T. et al., 1982. J. Bacteriology 150:269–276.

Yang, F., Moss, L.G., Phillips, Jr., 1996. Nature Biotechnology 14:1246–1251.

Ethical Dilemmas in the Conservation of Biodiversity

Anil K. Gupta
SRISTI, Indian Institute of Management, Ahmedabad, India

Introduction

The green revolution caused the maximum erosion of genetic diversity and that, too, under public direction and approval? Does it mean that when erosion of biological diversity of various species take place to ensure food security or ensure survival of humanity itself, it is justified? Otherwise not. Who will decide what is the correct order of preferences in this regard? If consumers want to pursue some tastes which require developing transgenic crops, will consequent risk on diversity be weighed only through pervasiveness of consequences?

Are moral dilemma then a function of scale of risk and not risk per se? The same problem choice of a small scale may be moral but on a large scale pose a dilemma. Perhaps, it is not the scale as much as the magnitude of manipulation of biological integrity of a specie or an ecosystem. If a new species of mouse is developed with a cancer gene, it is moral in so far as it helps cure cancer. Is it moral to do research on it but immoral to patent it or profit from its commercial use? It is moral to manipulate but not to benefit from it in such a way that integrity of a specie becomes a controllable resource. That is, one individual or group of people can own a species. The objection then seems to be to the idea of owning a specie and not changing its constitution unless it poses a health or ecological hazard. But then application of pesticides causes more hazard to soil microbial diversity, as well as to the health of farm workers in the Third World, but its application does not seem to raise as many moral dilemmas. Does the degree of concern have to do with the order of organism? The higher the form, the greater are the entitlements? The apes or dolphins have higher order entitlements but rats have lesser rights than cats (because the latter are reared as pets and rats are not). Will the rights of rats change if some people rear them as pets?

In this paper in Part One I deal first with the context in which regions of high biodiversity often have a high level of poverty. I review some of the ethical guidelines discussed in literature drawing upon my earlier work in Part Two (Gupta, 1994a, b). The ethical guidelines developed by Pew Conservation Scholars are discussed in Part Three. With the help of two cases of interface between biotechnology and biodiversity, we discuss ethical issues involved in transforming biodiversity for various purposes in Part Four.

Conserving Biodiversity Without Keeping People Poor

It has been argued earlier (Gupta, 1991a, b; SRISTI, 1993) that regions with high biodiversity also have high levels of poverty. The reasons are fairly obvious: one of the conditions which favors high diversity is high ecological heterogeneity at short distances due to changes in topography, drainage, or slope. The coevolution of biological and cultural diversity is well understood. However, what is less appreciated are the reasons that generate low purchasing power in the hands of communities living in these regions.

Because of diversity and a seasonally and spatially fluctuating supply of different natural goods, lack of access to modern means of communication, and demand generation, the consumer demand for these goods is very low. Further, the dearth of local employment opportunities because of low productivity and high risk creates pressure for emigration (generally males, but often entire families). The gender and age composition of the population makes the conditions for economic subsistence even harsher. The dropout rate of Indian children from school is the highest in these regions. It is not that in biodiversity rich regions local communities are poor in all resources. There is one resource in which people in these areas are extremely rich and that is their knowledge about biodiversity and its uses.

There are several threats to the diversity as well as to the knowledge which generates ethical as well as socioeconomic dilemma—

- If diversity has been reduced in the regions which have experienced economic affluence, should the remaining regions also go through the same experience?
- If erosion of knowledge continues to take place at the current rate, will biodiversity in the future be like a library of which the catalogue has been lost?

- If livelihood opportunities for local communities do not improve by conserving diversity, will they not be justified in reducing diversity to improve their options?

  Alternatively, should one try to conserve diversity by keeping people poor?

- If many of the local herbalists, healers, and other biodiversity experts are poor despite having a richness of knowledge but do not charge for their services, should they be kept poor because their ethical values are superior?

  Can their ethics of sharing be reciprocated by the consumerism society in a different way?

- Among the threats to diversity how do we assign weights to various threats such as deforestation, urbanization, excessive mining of ground water, mono cultures, disruption of natural corridors for wild life, introduction of genetically modified organisms, use of chemical pesticide and other inputs in agriculture, forestry, and other such areas?

- Should perception of threat to biodiversity be influenced by the potential hazards in terms of species, habitat, time frame (immediately or in the future), and reversibility of change.

- How should the threat to food security be weighed vis-a-vis threats to the other life support systems (for instance, if a transgenic crop eliminates the use of pesticides and makes the food production system more sustainable, is it preferred over greater damage by pesticides but with no risk of horizontal gene diffusion from transgenic to other plants?).

  Similarly, how should availability of a drug at a low cost to a large number of people by producing the same through transgenic plants be compared with potential death or disability of the poor people who cannot afford to buy the drug manufactured through current technologies without associated threat of gene diffusion?

There are many more questions which can be raised with regard to conservation, development, and reciprocity among those who conserve or develop or both. In the next part we discuss a framework which makes it possible to trade off these issues and thereby arrive at some principles for making decisions.

Conceptual Framework for Resolving Ethical Dilemma

Responsibility towards nature may arise from several kinds of motivation and likewise be tempered by various kinds of constraints. The motivators may include attributes like reciprocal or nonreciprocal altruism, concern for nature for its own sake, option value in terms of potential future gains through the use of diversity, critical vulnerability due to upstream-downstream linkages, respect for rights of different species to exist, day to day livelihood dependence, global willingness to pay for local conservation of endemic and endangered species whether in wild or domesticated sectors, religious or sacred beliefs, such as generally associated with the sites of origin of various rivers and mountain peaks.

The constraints for conservation may arise on account of (a) limited knowledge about a specie or a habitat or its interaction with other species, (b) lack of measurability of the occurrence or the rate of decline, (c) lack of constituency, particularly for less known species or habitats, (d) unwillingness of consumers to pay a premium on organic products or biodiversity based products, (e) lack of willingness on the part of drug, dye, or agrochemical industry to share the benefits with the communities or individuals conserving diversity and providing knowledge about its uses, (f) lack of poverty alleviation forcing local people to extract resources in a nonsustainable manner, (g) lack of regulatory mechanisms for external corporate or other biodiversity extractors from within or outside the country, (h) perverse incentives for excessive extraction of natural resources, and (i) lack of accountability of biodiversity users towards long-term conservation of resources.

The motivating as well as inhibiting factors interact within the public policy and sociocultural context. A society may put high premium on conservation of tigers or lions but may ignore conservation of frogs or snakes. Similarly, public policy may subsidize pesticides and excessive mining of ground water and pollution of surface water bodies but may not provide incentives for R&D in adding value to nontimber forest products. The outcome of this interaction shapes the preferences of conservators as well as users of biodiversity about technology domain, intensity, and purpose of use. Each one of these has implication for sustainability of livelihoods as well as biodiversity conservation strategies.

The framework provides the influencing factors as well as the impact points. The regulation of access and negotiation of terms of exchange can be guided by the institutional capacity of society and the dominant ethical paradigm. Several of the Pew Conservation Scholars discussed the seven issues based in a background paper along with other information to evolve ethical guidelines for accessing/exploring biodiversity. It was recognized in these guidelines that the relationship between outside explorers and local communities conserving resources could be guided by four possible modes of relationships, such as (a) commercial extractive, (b) commercial nonextractive, (c) noncommercial extractive and (d) noncommercial nonextractive.

The first category includes all attempts by local or external users who extract biodiversity resource for commercial purposes. These could include small-scale collection by local herablists for possible commercial returns as well as large-scale collection by external corporations or government departments for sale in the market. The ethical responsibility of extractors in this category obviously cannot be the same as that of scholars or academics or local healers who collect it for noncommercial purposes. A cytologist or plant physiologist may collect small samples for scientific study of diversity. The ethnobotanists who do not extract plant samples as much as the knowledge of people should obviously be regulated in their efforts so as not to rob the people of their only resource–knowledge. The ethical responsibility of those who neither extract any resource nor have any commercial purpose but merely try to understand the interrelationship between different species will invariably be much less than the first case. It is true that any knowledge even of basic ecological principles can be put to economic use. However, that generally may not be the purpose of an ecologist.

Granting the differences in the degrees to which different resource extractors should be considered bound by following ethical guidelines will mean the need for flexibility and practical judgments in the enforcement of guidelines. So far as the sustainability of extraction is concerned, the proposed principles could be the following:

- We should collect as much from as many places as can be taken without jeopardizing the renewability of a resource keeping margin for unforeseen contingencies and stresses. Assume that a particular herb or grass is found scattered in different habitats but at one place there are no more than 10 or 15 clumps. Depending upon the flowering, ratio of seed setting, and dormancy. one will have to decide what is the minimum strength of clump necessary in a habitat to renew the population under normal circumstances. Assuming that there could be a drought or excessive rain, some margin should be provided in the form of redundancy. The requirements of faunal diversity should be taken into account, and only then, the net availability should be estimated.

- The interaction of species should be closely monitored particularly when reduction in one may provide a niche for expansion of the other. Depending upon the nature of invasive species in a given habitat, the degree of extraction and its duration will need to be tempered. For instance, if nonedible woody species invade a pasture and become dominant, not allowing the edible herbs and crops to come up, then the quality of pasture may be irreversibly damaged unless the dominant specie is physically removed or reduced.

- Certain species of plants and animals play a crucial role in the sociocultural life of the local community. Elimination of access to these species may disrupt the rituals and other cultural needs requiring those species. The sustainability of a local cultural institution should be woven together with the need for maintaining the ecological integrity of a habitat. Otherwise, durable stakes of a local community may not develop in safeguarding a diverse resource.

- In case of endangered species, unless a full fledged program is in place for a short to medium term conservation of the species, the extraction should not be permitted. This should be a general guidelines for all kinds of extraction but should become obligatory in the case of endangered species.

- Many of the habitats of endangered species particularly of wildlife may also have separate sites or spots revered by local communities. Right of visitation to such sites in specific seasons or times may be necessary for a manageable number of people.

- The rate of extraction of various resources may sometimes require constant monitoring because of a lack of information about the potential ecological stability. One must remember that sustainability in certain cases require sustained extraction of resources. For instance, in Shivalik Foothills and Himalayan regions, if Bhabar grass is not removed from the hill slopes and forests prior to summer, the fire hazards increase enormously because the dried grass during the summer produces fire which engulfs large areas

> because of strong winds. Similarly, desilting in a tank becomes most necessary to maintain the capacity of the water body. Likewise, in certain bird sanctuaries limited grazing by animals may be very efficient and productive so that the grass does not become too tall, preventing birds from finding prey. The norms of extraction should, therefore, look at the need of removal of a certain biomass, as much as for leaving it in its place when necessary.

There may be several other factors which influence maximum sustainable yield from any ecosystem. However, recognizing that disturbances in the ecosystems are becoming more widespread and instability, in general, is on the rise, some allowance must be made for uncertain contingencies while calculating the maximum sustainable yield of resource. Identification of ecological indicators which give an early warning indication of potential threats to a diverse system may thus be an important prerequisite. People's knowledge of ecological indicators, interspecies interactions, and other technological dimensions mentioned above may be very crucial for sustainable extraction. Ethical concern will emerge while dealing with the use of this knowledge in a fair manner.

Ethical Guidelines

Guidelines by definition are normative indicators. At the same time, they provide an understanding of what seems right or reasonable, in general. In a specific case, guidelines will have to be interpreted in the spirit of mutual respect and accountability.

The values and the processes involved in the production of knowledge in the formal organized sector and informal colleges of peasants may be quite different in many cases. The production, reproduction of knowledge, its communication within the community and to outside professionals guided by scholarly motives or commercial concerns involves resolution of a large number of ethical and moral dilemma.

How do we treat the oral traditions which are likely to die away if patterns of social interactions continue to change the way they are changing at the present? What should be the incentive for scholars to document knowledge under such threats of erosion? What happens to the identity, authorship and proprietary right of the individuals, families and the communities which share this knowledge in good faith with outsiders? How should a corporation, whether in a public or private sector, national or multinational view access to documented or undocumented knowledge? Does the act of documentation of such strategic knowledge deny the provider of information of any right in the value added products or income therefrom? Once the providers come to know of the asymmetry in the returns accruing from the knowledge they provided to outsiders, could they become secretive? Will such a tendency not prevent future generations from getting access to valuable knowledge that might get lost because of the social pressures, migration, economic deprivation, and lack of young people acquiring the traditional knowledge skills? How do we balance the right of the present generation to benefit from this knowledge and underlying diversity with the rights of future generations?

The answers to these questions are not easy. The situation becomes even more complex once we include the microbial and aquatic diversity. The relationship between the human contribution and natural factors in maintaining or generating microbial diversity is not even properly understood. Similarly, avion, as well as aquatic diversity, may transcend boundaries of several political nationalities just as the plant biodiversity may do.

The movement of animals, whether on land, water, or in air from one region to another makes the claim and responsibility of different countries very complex. Fish found in waters of one country may go for spawning upstream in another country. If spawning grounds are not conserved, the fish diversity is unlikely to be conserved. How should such inter country agreements be drawn up if industry, based on such fish, is in downstream country? Apart from the legal claims, respect and responsibility for life, particularly in its reproductive phase, have been a part of culture and values of many societies. There are institutions which prevent people from fishing during the spawning period. Such institutions did not necessarily emerge to balance claims and counter claims of the conservator of biodiversity and its uses. However, in a world dominated by economic calculus, value based institutions are coming under strain.

Similar is the situation of the communities which conserve biodiversity but refuse to receive compensation. How should the society view its responsibility towards such communities and individuals? Do we exploit them because of their superior ethics, and accumulate wealth at our end? If not, then what kind of moral and legal instruments and incentives can be generated which do not force people who conserve biodiversity to remain poor? The situation is complicated by the fact that the time and resources that a corporation may put in identifying and developing a product based on local biodiversity may be more easily computed. And thus, this amount may appear very large. The value of the time and effort spent by a community in safeguarding local biodiversity by a combination of ethical and moral principles and institutional practices is difficult to work out. The opportunity cost of their time, as evaluated by the market, can obviously not be a proper indicator. In many cases, the contribution has been made over centuries.

Some colleagues (Nietschman and Churcher, 1994) have tried to compare the existence and conservation of biodiversity with the presence and exploitation of mineral resources like gold or oil. Not realizing that in existence of gold or its conservation underground, human intervention was minimal (of course, by not exploiting it, it could just as well be conserved), the comparison has problems. No human being could reproduce gold. The biodiversity, on the other hand, can reproduce differently with or without human intervention. To argue that claim of a society or a community accrues entirely on account of an ecohistorical or geomorphic reason without any conscious contribution of a local knowledge system and its institutions will be to miss an important point. And that is the selection pressure that human beings have put discriminatingly on a given biodiverse system. Whether it is through fires in the forest, restriction on fishing in certain seasons or regions or through conservation of certain weeds or utilization of certain wild plants, communities as well as knowledgeable individuals have contributed a great deal to the conservation of biodiversity. Similarly, contribution of a state cannot be ignored altogether. By imposing locational costs on the enterprises for which forest land under government possession may or may not be transferred, the cost of conservation is partly internalized by the enterprise and partly by the society through higher product costs.

The issue is not whether intellectual property rights fully capture the contribution of local communities and individuals, given the complexity in conservation. The issue is whether IPRs of local innovators, communities, and traditional herbalists do help the outsider users of this knowledge resolve their own moral and ethical dilemma.

It is one thing to say that locational advantages of local communities must not be given undue weight. It is quite another to suggest that outsiders should follow different ethical principles while dealing with knowledge produced by their colleagues in labs and libraries and knowledge produced by people over time, individually or collectively.

To argue that the problems of communities which are in conflict with the nation states of which they may be a part of at the present time must be resolved first and only then the issues of compensation should be addressed is to bypass totally the issue of our own accountability.

In the brief review of some of the important studies and documents, various principles which have been identified by various institutions and scholars, though the same may not have been implemented even by the organizations or other individuals espousing these principles, are summarized in Part One of the paper. In Part Two I present the discussion on ethical guidelines developed by a group of Pew Conservation Scholars based on several studies by us and other colleagues (Gupta, 1994a, 1994b; Nietschman and Churcher, 1994).

Review of Ethical Concerns

The World Health Organization (WHO) in collaboration with International Union of Conservation and Nature (IUCN) and the World Wildlife Fund (WWF) issued the guidelines in 1993 based on various consultations on the subject held earlier. The guidelines conformed to the principles stated in Caring for the Earth, a document prepared by IUCN, United Nations Environmental Program (UN-EP) and WWF in 1991. It may be useful to add that the earliest reference in this document pertained to 1979 and 90 per cent of the references were post 1985 and western in origin. The inference was obvious that there was nothing in the various religious, cultural and ethnic traditions of Hindus, Muslims, Masais, Buddhist, Inuits, Akwasasne, or Zuni which had anything to contribute to the evolution of global world view (Gupta, 1991, 1992). It is such an assumption which has guided the development of guidelines by various international institutions–particularly those involved with IUCN and WWF. The WWF guidelines emphasized among other issues, the following principles:

- The ethnobotanical data should be catalogued and analyzed but disseminated in such a way that data providers receive benefits from the commercialization of the product based on the information.
- The Ministry of Health should incorporate proven traditional remedies into national programs of primary health care.
- The traditional health practitioners should constitute themselves into a national body.

The Chiang Mai Declaration To Save The Plants That Save Lives issued in 1988 (basis of these guidelines) was certainly a step forward. However, the responsibility of international organizations and private national and multinational corporations which had drawn upon the indigenous knowledge for so long was not clearly defined. Similarly, no attempt was made to identify institutional mechanisms for ensuring compensation to the communities from whom resources may have been taken. It was, in fact, acknowledged (p. 14) in a footnote that under the given laws, most communities did not have a legal right to their traditional knowledge. Further, in most cases, the communities could not be identified in the legal sense.

The only problem with this view is that a whole community may not have a legal right, but the right of a corporation adding value to a knowledge provided by a medicine man or woman living in a community could not be superior to the right of the community. It is this asymmetry which these guidelines have chosen to ignore.

Ethics, Ethnobiological Research, and Biodiversity

The WWF (1993) brought out in the same year another report on the above topic written by Cunningham. In this report, it was recognized that monetary and nonmonetary utilitarian values justified the conservation of biodiversity. The question of the fair share of benefits for the providers of biodiversity with or without associated knowledge was raised. The guidelines developed for ethnobiological research underlined the following principles or procedures:

- A strict code of professional ethics to ensure that research participants and members of local organizations are informed of the objective, commercial aspect, and possible results of research; confidential information and request for anonymity is respected; equitable compensation for assistance provided by the individuals; and fair royalty payment to regional/national organizations and acceptance of national requirements for prospecting biodiversity.

- The explorations should involve local people in screening. A commitment should be made to transfer technology and to provide training in cataloguing and screening biodiversity with government assistance.

- Supply agreements should be made with reputed organizations and not with individuals who could be guided by personal gains.

The report acknowledged that the information obtained by the ethnobotanists through a relationship with local communities based on trust is used for the publications and commercialization. The respect for local rituals was often diluted. Despite the fact that the hit rate was much higher with ethnobotanical collections compared to random collections, the corporations and governments around the world had not pushed for any fundamental shift in the responsibility of ethnobotanists.

For instance, even this report did not acknowledge the unethical practice of claiming authorship for knowledge by the ethnobotanists, who merely chronicled the information provided by others. These, "others" remained nameless and faceless even if the community was acknowledged.

The Declaration of Belem in Brazil recognized the need for compensating the providers of knowledge and also the need for ethnobiologists to share the results of the research with native people in their native language. However, the authorship protocol was not modified. Even though the author (Gupta, 1992) had made a personal oral appeal to the President of this congress prior to the congress (in Bad Ball, Germany, 1992) no resolution was passed by the international congress, insisting that if the researchers did not share their knowledge with the providers, their papers would not be accepted for presentation in the next congress. Similarly, no decision was taken on acknowledging providers of knowledge by name.

National Cancer Institute Guidelines

The U.S. National Cancer Institute guidelines (1991) have been more advanced than the Belem Declaration. It called for the protection of IPRs, compensation for traditional knowledge, though not necessarily in the form of cash, responsibility of multinational corporations and academics involved in the use of natural products, and traditional knowledge to recognize the responsibility in compensation and provision of intermediate compensation until the final product was developed and the profits earned.

The amount involved was 2.7 million dollars in 1986 and 3.8 million dollars in 1991 in three 5-year contracts. Obviously this amount may appear abysmally low compared to the potential value of the information, and only a very small share may have gone for actual payments to the people to improve their lives or for local infrastructural development.

Guidelines for Foreign Collectors

In another workshop in Australia, 1990 (WWF, 1993:20) the guidelines for foreign collectors required collectors to pursue the following obligations: (a) to arrange to work with local scientists and institutions, (b) to respect regulations of the host country, (c) to obtain official permission, (d) to include travel expenses and the cost of other activities incurred by the host institutions, (e) to leave behind a complete set of properly labeled duplicates before leaving the country, (f) to inform the results in the country of origin, (g) not to exploit natural resources in an unauthorized manner, (h) not to violate sanctions against collection of endangered species, and (i) to send copies of the research papers and reports to the host institutions and collaborators after acknowledging their contribution. Even these guidelines did imply a much greater accountability on the part of expatriate researchers though the IPR and compensation issues were ignored.

UNEP Guidelines for Country Studies on Biological Diversity

These guidelines, developed to assist countries in assessing the status and value of their biodiversity, are important reference points for any discussion on ethical and value issues. When a document as important as this does not emphasize the responsibility of organizations collecting data towards the providers of knowledge and/or conservators of biodiversity, the lapse becomes noteworthy. Twenty guidelines, included in this document, do not refer to the norms that should guide the accountability of data collectors and users towards the providers of information. It also does not take care of the cultural values of the people that may be crucial for the conservation and may provide clues to the kind of relationship a community may like to have with outside institutions. The technical annexures to these guidelines included a detailed section on socioeconomic factors affecting biodiversity. There is an acknowledgement of the fact that property rights may vary across and within a resource (p. A–3) but these rights refer to physical resources. The cultural factors are incorporated here–more as a description of factors influencing conservation rather than as a protocol governing relationships with the community. The section on biodiversity services includes a whole range of services, such as genetic material conservation, and watershed management, but the existing contribution of local communities to the provision of these services is not specifically recognized. The section on economic value of biological resources and biodiversity (p. C–54) refers to different kinds of use values. Even here, the value contributed by the knowledge systems and, thus, in some cases may be returned for exclusive use of a small group would not be captured because of the prevalent norms of a community.

In some cases the community may insist that exchange of a particular resource must not be attached to a commercial value and, therefore, it might be characterized as nonmonetary use. On the other hand, restrictions on collection of plants through cultural means may determine the supply and consequent costs for the users. There is a need for modifying the concept of existence values as well as option values. The existence values incorporate the ethical belief and presuppositions, whereas option value refers to future uses possible through prospecting biodiversity. The existence value can be assigned only if the ethical and moral principles guiding the conservation are considered legitimate and consequential for national as well as international policies. The value that society is willing to pay to ensure future access is option value and would vary across different political and social regimes. The guidelines in this section do refer to these values, but the methodology and examples provided don't seem to do justice to this concept.

Guidelines of the American Society of Economic Botany

The International Society of Economic Botany (Baba, 1993) has developed draft guidelines on professional ethics. The guidelines clearly provide what we have been pleading for in the last 5 years that the "members of the society for economic botany have responsibilities to those studied":

- They will communicate clearly and honestly to all informants, the objectives and possible consequences of ones' research. If the research has a commercial objective, they will make that explicit and will disclose what the commercial results might reasonably be expected to be.

- They will comply with all rules and limitations informants and institutions place on the research. They will not "trick" informants into revealing "secret" information. They will supply any reports or results that are requested.

- They will respect any request for confidence made by those providing data or materials, provided that the maintenance of such confidence does not compromise other ethical considerations.

- They will respect informants' right to anonymity and privacy when requested.

When materials or information obtained from informants can reasonably be expected to have a commercial payoff, they will arrange with employers for equitable economic compensation for the informant(s) and will do all in their power to ensure that compensation is paid.

These guidelines are in some respects more precise and advanced than the Belem Declaration. However, it may be useful to clarify that the research results should be shared in the local language and be comprehensible, regardless of request. These results can be deposited in local schools or counsel offices, apart from posting these to some of the informants. When requested, these must go to all who request as per the guidelines.

The observation in the guidelines about responsibility of the economic botanists to do all in their power to ensure that the compensation is paid, is one of the most forthright statements expressed by any professional body so far. The idea (4.B) that "they will not present as their own, the work of others," can be made more explicit to suggest that whenever informants do not prohibit, their identity will be acknowledged as a part of the research paper just as personal communications from fellow colleagues are acknowledged in academic writings.

One of the major weaknesses of most declarations and guidelines is that these acknowledge the unjust manner in which biodiversity had been exploited in the past and yet did not suggest concrete obligations that the national and multinational corporations should bear. Neither have these guidelines distinguished the degree of responsibility with the nature of purpose.

Khalil, Reid, and Juma (1992) in a review of the conference proceedings on Property Rights, Biotechnology, and Genetic Resources argue that apart from code of conduct and state recognition of individual and community rights there was a need to develop new IPR mechanisms building upon the idea of the "corporate" community. Hecht and Cockborn (1989), in an appendix, provide a forest peoples' manifesto in the context of Amazonian development and insist the people living in the forest should be allowed the right to participate in public discussion (through their representatives) in regard to all government projects for forests inhabited by Indians and rubber tappers as well as other extractive populations.

National Science Foundation guidelines to apply for grants for research and education in science and engineering expect that the findings of the research and educational activities supported by NSF be promptly shared with other researchers in a reasonable period of time through publications with proper acknowledgment. However, it does not require that the research finding be shared with the providers of the data. Similarly, guidelines allow awardees to retain principle legal rights to intellectual property but say nothing about the rights of the communities or individuals who may have provided the valuable clues.

The Biodiversity Convention office of Environment Canada brought out a document entitled, "Biodiversity: The Web of Life (1993)." An indication is given that Canada would support a further round of negotiations on global trade after the Uruguay Round of GATT has been finalized–the environment being a focal point. This is one of the rare proclamations from a developed country on the subject. How far this will help in improving accountability towards local communities remains to be seen.

Rafi has been a persistent campaigner on the subject of patent rights of people. In general, Rafi has been opposed to any kind of patents on life forms seeds. However, in a recent public debate on the subject held in Canada, Pat Mooney questioned the possibility of Third World people getting any advantage from the intellectual property rights regime and argued for continued free exchange of germ plasm. The author (Gupta) argued that such a view implied that the contemporary innovations in the developing world would not be able to pursue their claims in the current international intellectual property rights regime. It is true that the legal costs and complexity may be beyond the current capacity of local communities in most developing countries. At the same time, that is not a reason for denying people their rights.

The report on conserving biodiversity by National Research Council (1992) acknowledges the need for greater emphasis to be put by development agencies on local knowledge but says nothing about the ethical values that should guide the conduct of development agencies or researchers.

Baenziger, Kleese, and Barnes (1993) in a special issue of Crop Science Society of America (CSSA), raised several other ethical and social issues in the context of IPR. The material transfer agreements have been proposed as tools to promote exchange of germ plasm among different nations and parties. The issue raised is, whether Material Transfer Agreements (MTA)s are appropriate instruments for conservation of indigenous genetic resources. What would be the appropriate forms of compensation in lieu of accessing genetic resources of a developing country–cash, debt write off, or R&D infrastructure? They also question whether the external cost of protecting or not protecting genetic resources of the Third World are known. On the other hand, they enquire whether use of MTAs would increase the cost of research and development thereby limiting the rate of technology progress. Are the restrictions sought to be imposed on the sale of farmers saved seeds under the UPOV 1991 justified? Should all the farmers be punished, they ask, for transgressions of a few?

The authors in the CSSA document recommend that intellectual property laws for plants should be rewritten and should apply to all kinds of plant intellectual property. The cheaper conflict resource mechanisms should be developed so that institutions which can't afford costly court fees can also claim justice. They argue for compensation to developing countries but do not discuss the issue of rights of communities and individuals within the developing countries.

Hamilton, Ellis, and Levitt (1994) review the issue of IPRs and refer to the comments made by Peter Day, Chair of the Committee on Agriculture set up by the National Research Council of the U.S. Day while presenting the report on Managing Global Genetic Resources (1993), offers the following three solutions to resolve north-south conflict—

- A new treaty should be negotiated which would define a compromise position on IPR as well as free flow of crop germ plasm (the Biodiversity Treaty did include some of these aspects but it has not been ratified by the US Congress as yet).

- An international payment system should be created and linked to seed sales. Proceeds could support genetic diversity conservation programs (similar to the International Gene Fund recommended by the Foreign Agricultural Organization (FAO) Undertaking on Plant Genetic Resources, 1993).

- Attention should be focused on increasing the capacity of developing countries to do research on plant breeding and other biological aspects rather than on legal arrangements. This would assume a kind of reciprocity between the industries of developing and developed countries. They note that American universities spent over 37 million dollars on legal fees to claim the IPRs of which only 12 million dollars was reimbursed through various licensing arrangements. This is out of 172 million dollars as income from licensing of inventions (one has to think about the capabilities of developing countries to bear such costs to enforce their rights in developed

> countries). The authors seem to share the general disapproval of the very broad patent granted to Agracetus for transgenic cotton. The authors conclude that the U.S. President's recommendation to the U.S. Senate urging ratification of the biodiversity convention is an acknowledgement of the basic principles of rights and responsibilities in this regard underlying the convention. They also foresee that debate on these issues is going to take time to be resolved.

The Northern Affairs Programme of the Department of Indian and Northern Affairs of Canada developed some guidelines for responsible research. It provided that the conductor should address the issue of intellectual property right for the traditional ecological knowledge collected from indigenous people.

The Draft International Code of Conduct for Plant Germ Plasm Collection and Transfer (1991) looked into the issue of the accountability of plant collectors and their ethical behavior. The code is voluntary and primarily addressed to the government and international organizations. The code aims to maximize the benefit to the international community and minimize adverse effect on evolution of crop plant diversity and the environment. The national governments were empowered to issue a license for collecting germ plasm as per their legal requirements. The collectors were expected to "respect local customs, traditions, and values and should demonstrate a sense of gratitude and reciprocity towards local communities.... The acquisition of germ plasm should not deplete the populations of the farmers' planting stocks or wild species.... While collecting cultivated or wild genetic resources, it is desirable that farming communities and caretakers of such resources be informed about the purpose of the mission and about how and where they could request and obtain samples of the collected germ plasm.... A duplicate set of all collections will be deposited in the host country."

As is obvious from the voluntary code, there is no mechanism for enforcing these guidelines. Neither is there any evidence of the code having influenced the behavior of collectors in any significant manner in the recent times.

Alcorn (1992) made several suggestions on the role of ethnobiologists and particularly pleaded that the others' knowledge should not be packaged as one's own work. She recalls the pioneering work of the Mexican ethnobiologists such as Toledo, Gomez—Pompa and Xolocotzl who asked the question about ethnobotany for whom. The report on the State of the Peoples brought out by Cultural Survival (1983) include a draft declaration on the rights of indigenous people. These rights pertain to all aspects of relationships between indigenous people, their culture, access to natural resources, their aspirations and their institutions. The collective intellectual property rights to the biodiversity and knowledge about it are also affirmed. In addition the right of free and informed consent through their representative organizations is also stressed.

The guidelines on Community Based Public Health Research Principles and Application Procedures adopted by Detroit-Genesee County Community and the University of Michigan, School of Public Health, (1994) recognized the right of community to be informed of the project objectives, procedures and findings in clear language, "respectful to the community and in ways which will be useful to the community." Any publication resulting from the research was expected to acknowledge the contribution prior to submission of material for publication and involvement of local collaborators as coauthors.

In some ways above guidelines in combination with the guidelines of the Society of Economic Botany do register an advance in our thinking.

Chapman (1993) in the agenda note for a conference sponsored by American Association for the Advancement of Science asserted that intellectual property rights of the indigenous people were a collective right held by virtue of membership in the group. He added that like any other collective right, "they pertain to protection of nonreducible collective interest and to promotion of common good." He refers to the revised draft (1993) of the universal declaration of indigenous rights first issued in 1991 by the Working Group on indigenous population. "The right to special measures for protection," the document notes, "as intellectual property of cultural manifestations, seeds, genetic resources, medicine, or knowledge of the useful properties of fauna and flora." The revised text Article 29 reads

> "They have the right to special measures to control, develop and protect their sciences, technologies, and cultural mechanisms, including human and other genetic resources, seeds, medicines, knowledge of the properties of fauna and flora, oral traditions, literature, designs and visual and performing arts."

This declaration, of course, has not been approved by many governments because the right of peoples might imply political independence.

Hamilton (1993) in a comprehensive study on ownership of plant genes focuses attention to some of the crucial issues in the debate on conservation of diversity and responsibility towards those in the Third World who actually conserve it. Many in the Third World question the validity of TRIPS provisions under GATT which did not provide any protection for unmodified plant genetic resources. On the other hand, the argument is that by putting a value on plant genetic resources, mechanisms for financing their conservation may be developed more easily. In my view, keeping the free access to germ plasm did not bring any benefit to the communities which conserved these resources so long. Maybe putting a value will. Hamilton questions whether Thomas Jefferson, who wrote the first patent law of the U.S, would have permitted the person who "discovered" the plant or the scientists who "engineered" the gene to be granted a legal right to own it.

Maybe not. So?

The Convention on Biological Diversity (CBD) obliges the scientific community as well as commercial extractors of biodiversity to follow sustainable methods of extraction without endangering the diversity and the knowledge about it. CBD further requires that the knowledge evolved by innovative individuals and communities should be used through approval and involvement of the communities and individuals ensuring equitable sharing of benefits. The concept of prior informed consent requires that consenting parties be made aware of the potential value or worth of what they are sharing with outsiders before entering into any transaction. However, as per the CBD, this consent is required only from the countries signing the treaty and not necessarily the communities within the country. No legal framework, no matter how comprehensive, can ever substitute the need for ethical guidelines that govern the behavior of various stake holders. Several Pew Conservation Scholars got together three years ago to make an attempt for developing such guidelines. Three background papers provided the context for the discussion and the author helped in steering the discussions so that a consensus statement could be prepared. Among the seven issues that were identified only about four could be taken up for detailed discussion and development of guidelines. I discuss the context, the background issues, and the guidelines evolved by the group so that readers might consider how these guidelines could be operationalized in different cultural and professional context. It is obvious that different institutional contexts will require same guidelines to be operationalized differently. However, there is a debate on desirability of such guidelines by various professional associations of breeders, botanists, industry and scholars involved in herbal drug development, and ethnobiologists. The experience so far has not been very encouraging. The professional bodies which have developed guidelines have to generate consensus for internalizing these guidelines so that the quality of professional decision making reflects the spirit more and more.

Guidelines

The earlier research has shown that regions of high biodiversity are often inhabited by people considered to be economically poor by international standards (Gupta, 1991). It was also realized while developing guidelines that if economic conditions did not improve, the biodiversity could disappear. We have to ensure that cultural, social and economic context of biodiversity is modified in such a way that local communities have a greater say in determination of the how, how much, where, when, and for what purpose, with regard to the conservation and utilization of diversity. Several principles drawn from earlier experience of participants and the spirit of Honey Bee Network were considered as fundamental to development of these guidelines:

- Research is an educational process leading to mutual earning among researchers and the collaborating individuals, communities and institutions.
- Just as the proprietary rights of scientific knowledge are well established and respected, such rights are due to the producers and providers of traditional knowledge and contemporary innovations from local communities.
- Research should be based on respect for the local cultural values and norms.
- Benefits should accrue to all partners in a fair and equitable manner.

It is obvious that mutual learning among communities and scholars is only possible if discourse takes place in the local language. Further, just as citation norms are well established in scholarly discourse, these are to be developed and respected while dealing with the knowledge of local communities and individuals. Similarly, one cannot usurp the knowledge of people without proper reciprocity and equitable sharing of benefits. Honey Bee Network articulated these values eight years ago when there was no consensus even among the social conscious scholars on this subject. The Honey Bee Network was based on essentially two attributes of the honey bee that outsiders could emulate in their own conduct: 1) the honey bee collects pollen from the flowers without impoverishing them, and flowers don't complain, and 2) it connects one flower with another through pollination. In addition, it also extracts honey. When we (outsiders) collect knowledge of people without any reciprocity, people can and should complain. When we do not share what we learned with the people in their language, the question of accountability and ethical responsibility does not arise. The worst is the case of becoming authors of information provided by farmers. Imagine a scientific meeting in which secretaries who chronicle the discussions become the authors of the ideas discussed. We would consider this unthinkable. And yet, this has been going on in the field of ethnobiology. The sharing of benefits would take place only after we have resolved the problem of minimum acknowledgement.

The guidelines dealt with different kinds of research relationships. It was acknowledged that one could not expect similar kinds of constraints on those whose purpose is merely to describe the ecological systems–those who want to extract biodiversity and associated knowledge systems for commercial purposes. Four kinds of relationships were considered based on the presence or absence of extraction and commercial and noncommercial orientation, such as these:

> *Nonextractive noncommercial research*—biologists document the evolution of species and ecological patterns and processes through observation and simulation without collection of samples.
>
> *Extractive but primarily noncommercial research*—this might involve collection of samples of organisms for description or for analysis of the interrelationships among species.
>
> *Nonextractive research with possible commercial potential*—ethnobiologists may study plants and animals without collection of samples. These studies may involve documentation of local innovations, traditional knowledge and practices, development of data bases of such knowledge, publication of books, films, or other forms of dissemination of local knowledge–for instance, electronic communication and CDs. This local knowledge may be documented in order to preserve it or share it within the community or beyond it.
>
> *Extractive research intended for commercial development*—extraction could be in small quantities, such as for biotechnological laboratories, or in large quantities for natural product development. Such research done by students, academic researchers, corporate researchers, or local communities

may be intended to develop new products based on biodiversity traditionally used by local communities or elaborated by individual innovators. It may also involve screening and biodiversity without making any reference to local uses.

There could be several other variations of these relationships. For instance, nonextractive, noncommercial research may generate information about certain taxa which are likely to contain a particular compound in high measure. This information can then generate commercial returns. A good example is that of taxol. Once the discovery was made in the U.S. about its anticancer properties, search began for other locations where other species of the same genus were available. Accordingly, India provided a major resource. It is a different matter that it has been extracted in a nonsustainable manner. However, merely the possibility of future economic uses should not prevent us from allowing access to biodiversity for generating more scientific information. The necessary guarantees may, of course, be obtained.

Similarly the nonextractive but commercial kind of exploration of the knowledge of the people about biodiversity could lead to extraction of resources once the commercial use is upscaled. That again cannot be a reason for not documenting local knowledge. We have to recognize that the technological solutions for institutional problems will not work. Thus, whether knowledge will be used for sustainable or nonsustainable extraction is a function of institutions in place and not the kind of knowledge collected. However, the method and values involved in collecting knowledge can indeed influence the use of that knowledge. For instance, if the researcher collecting knowledge is obliged to take prior permission of the provider before sharing it with anyone else, then the provider has some control about the use to which it can be put. The guidelines dealt with primarily five issues: approval, initial disclosure of information, involvement and negotiation, compensation and other terms of access and responsibility of professional societies, and academic institutions and funding agencies.

Approval

In most cases the researchers should obtain clearances from appropriate central or state government authority and, where applicable, from institutions of indigenous people.

The problem may arise in countries where either division of responsibility is not clear or appropriate authorities have not been established. The lack of institutional structures should not be taken as an excuse for illegitimate access to biodiversity. How should one resolve the dilemma in such cases? A mullet national company developing herbal drugs faced this problem when it received unsolicited offers of sharing plant samples, along with their possible uses from a developing country that was otherwise reasonably strong in terms of its democratic institutions. The company wanted to follow the spirit of CBD and yet did not know whom to approach to find out whether such samples can be accepted without violating any national law of the country. The relevant ministry in the country concerned has not developed a policy and the business as usual continues. The company in question of course did not pursue the business in its anxiety to be ethical. The lack of appropriate institutional mechanisms may crowd out the ethical practices and thereby make it more and more difficult for those who voluntarily follow ethical ways of accessing diversity.

Initial Disclosure of Information

The guidelines provided that the researcher (while contacting a community or individual to seek access) should do the following things:

- Carry out all communications in the local language.
- Explain the nature and purpose of the proposed research, including its duration, the geographic area in which research would take place, and research and the collecting methods.
- Explain the foreseeable consequences of the research for resources, people, and accessors, including potential commercial value.
- Explain the potential noncommercial values, such as academic recognition and advancement for the researcher.
- Explain any social and/or cultural risks.
- Notify the community at large by some means; i. e., public meeting.
- Consider explaining the guidelines that the researcher is following, as well as his/her practice in previous similar research projects.

- Be willing to provide copies of relevant project documents, or summaries thereof, preferably including the project budget, in the local language. In the case of commercial prospecting, researchers must share such documents.
- Agree on a protocol of acknowledgements, citation, authorship, inventorship as applicable, either citing local innovators or conservators or respecting request for anonymity.
- Share findings at different stages with the providers.
- Not engage in bribery or making false promises.

In many cases the familiarity of a community with the organization of the researcher concerned may not require explanation of all the dimensions of one's involvement. However, this is the toughest part of the guidelines in which we find the hesitance among the researchers to be the highest. Apparently, the difficulty in sharing the project documents, compensation for scholarly pursuit and unwillingness to share one's own material gains with the communities seem to be at work. This is not surprising, given the history of relationship between outside academics and local communities. However, as said earlier, when knowledge of people is the commodity in exchange, then the contribution of the provider of knowledge could not be considered less valuable than the one adding value. In practice, perception will vary on the relative value of different contributions. However, the responsibility towards the provider cannot be disputed. Some of the professional associations specifically advise against any false assurance or commitment to the provider of information in order to gain access easily. The difference between, "must," "should," and "should consider" in these guidelines is intended to demonstrate the consensus of relative importance. It is quite possible that in different contexts some of the 'should may become 'must and vice versa. However, this is an issue on which consensus across cultures and professional institutions needs to be achieved, at least on the minimum core values.

Involvement/Negotiation

The guidelines required that in negotiations the researcher do the following things:

- Make a reasonable effort to identify and negotiate with those with the proper authority to negotiate.

- Conduct initial discussions with small groups (but obtain final approval from higher legitimate authority wherever applicable).
- Consider (where there is no existing authority or capacity for such negotiations) helping the community develop the institutional capacity to appraise and (if it chooses) enter into such agreements.
- Be willing to provide copies of relevant project documents, preferably including the project budget.
- Disclose commercial interest or other possible interest of present or future third parties.
- Include a local institution as partner in research, where an appropriate one exists.
- Consider drawing up a collaborative agreement.
- Consider depositing a copy of it (if an agreement is made) with a relevant regional/subregional body.
- Ensure that the actual entity directing the research is a party to the agreement whether they are carrying out the work themselves or through contractors.

One of the crucial issues here is the responsibility to develop the capacity of the local communities or the authorities concerned to negotiate and enter into informed agreements. It is well known that a large number of communities have an ethics which makes them share their knowledge without any expectation of reciprocities. For such people the concept of contract for sharing knowledge is totally alien to the existing world view. At the same time, one could not argue that their superior ethics should he responsible for keeping them poor indefinitely. Sooner rather than later, this knowledge will get eroded because younger people in the community may not have any incentive to acquire this knowledge. Similarly, one must accept the fact that a very large number of projects, designed and implemented by international as well as national institutions, have basic information about exploration, extraction, and benefit sharing that often is not widely shared. The guidelines suggest the need for greater transparency as well as responsibility. One of the dilemma which may arise in this regard is that local institutions may not give primacy to the interest of local communities while entering into different kinds of contact. In such situations researchers may have to arrive at multiple contacts to safeguard the interest of the communities without compromises with the law of the land. The involvement of the local institutions is important for institution building purposes. In other words, the renewability of the research process will depend upon how well the goals of capacity building have been pursued. For instance, if a majority of the training resources are so allocated that only the technical level competence is built in the local institutions, whereas, the scientific and scholarly capacity is augmented in external institutions. This may appear an expedient solution in the short run but in the long run may generate distrust and lack of proper reciprocity.

Compensation and Other Terms of Access

The researcher must do the following things:

- Make every effort to ensure that providing communities and counterpart institutions will share equitably in the benefits.
- Make every effort to develop effective mechanisms for benefit-sharing, (recognizing that no proven universal methods exist, and that cultural and other circumstances will vary widely from one case to the next).

Parties should arrive at the scope, extent, and form of compensation and keep all the following stages in mind:

- When accessing is done.
- When a new use is discovered.
- When a product is developed.
- When commercialization is done.

Arrangements for compensation should incorporate the following obligations:

- The community's right to any organism or part thereof extracted by any biotechnological or other method must not be exhausted merely by publication or collection. The community can assign these rights or associated intellectual property rights (IPRS) to anyone it feels appropriate.

- The community has the right to refuse collection by any researcher even after the initial research has shown its utility.
- Any research collecting from an alternative location, community, species, or country should take into account the contribution of the original source in generating commercial returns.
- The period of production should be considered to be valid as per the law in force for the property or form of accessed material being commercialized.
- At stage "b" or "c" above, researchers must negotiate with the source community the terms of profit-sharing from commercialization, even when knowledge is provided by an emigrant belonging to that community.
- Researchers should help to set up local community-managed institutional funds or other augmentative mechanisms for local community development in cases where individuals/communities refuse monetary compensation.

The need for graduated contracts may be a useful strategy to encourage interactions in a spirit of shared risk. The disadvantage is that the local communities may loose interest if commitment for upfront payments for conservation and economic development does not take place at all. The balancing of three goals of conservation product development and economic development is obviously not easy and may not take place in every case. Some scope for cross subsidization should exist and that is possible if contribution to some common funds maintained at regional or national is considered necessary by every party entering into any contract with local communities.

It is also true that a large number of nongovernment organizations (NGO)s and local groups raise expectations regarding immediate large payoffs so that not many contracts actually take place. Consequently, some other communities more pragmatic in this regard succeed in getting research contracts organized and with increase in the information flow try to revise them as per agreed milestones. We need to learn about this process by doing this in the best possible manner so that eventually good standards can be set. Given the information asymmetry, it is inevitable that it will require much greater responsibility on the part of dominant actors. The tradeoff is between shared exploration, moderate gain and higher competence, better negotiation and higher gains in the long run. We may reduce the prospects of exploration altogether by increasing barriers to negotiation.

Professional societies, academic institutions, and funding agencies should do as follows:

- Encourage citation of intellectual contributions of local innovators, communities and groups.
- Ensure sharing in the local language the insights gained from local communities or innovators either by the prior agreement or by the time of publication, or within reasonable time but not beyond one year of publication.
- Help set up a system of registration of innovations/practices so that IPRs of local communities or innovators are not exhausted.
- Set up rules of good conduct and practice by researchers.
- Recognize, support, and reward ethical practices in research.
- Set up bioethics committees to protect the rights of researchers, communities, and individuals contributing to the conservation of biodiversity.

The role of professional societies is most important.

If National Science Foundation Guidelines require that all assistance should be acknowledged adequately, it could also require that all informal sources of strategic communication (for instance, ethnomedical uses of plants) should be acknowledged, cited and provided a copy of research reports in an easily comprehensible manner. So far as professional conduct is concerned, it is the collective consensus in professional societies, academic institutions, and funding agencies which will help in achieving proper ethical responsibility towards communities and individuals conserving biodiversity and knowledge around it.

Blending Ethical Concerns With Biotechnological Potentials

Case 1: Should responsibility towards domestic/western diversity be different from Third World/global biodiversity? Hodges et al. (1996) provide an interesting case of herbicide resistant rice with a commentary by Comstock dealing with the dilemma of allowing research in one context but not another. Red rice is reportedly a serious weed in rice fields of North and South America. It is very difficult to control and crosses with rice approximately 2%—a proportion supposed to be very high by plan breeding standards. A herbicide resistant gene has been introduced into a rice variety. The international donor agency which funded the research did not allow this rice to be released for production in the U.S. due to the risk of gene transfer to the weed. The weed would have also become resistant to herbicide. Another scientist at an international research agency learned about this rice and got some seed for commercial production. His aim was to eliminate the chemical load and reduce production cost in Colombia where chemicals were used extensively for the purpose.

Among various issues arising from this case, Comstock (1996a) questions whether developing herbicidal resistant (HR) rice was such a good idea. He questions the rationale because it did not reflect the best way to control the weed. He prefers that scientists develop variety which were resistant or could out-compete weeds. He also wonders why something which is bad for the U.S. is good for Colombia? Finally he asks who should decide questions of this kind–the U.S. government, funding agency, international research agency, Colombian government, an international regulatory committee, the farmers, private business, or consumers?

Undoubtedly the case does present dilemma shared above but it also raises vital questions about the valid arena of judgments. It is quite possible that Colombian agricultural research agency in full cognizance of the risks involved might have asked for the seed in order to meet some short term food security goals. In fact, many developing countries are doing this. The question then is whether a legally valid transaction also becomes legitimate? Also how can will of sovereign nations be overridden even if it is not in the vital and long term self interest by an international regulatory regime? Perhaps the ethical issues involved cannot be resolved merely through more information but by proper weighing of pros and cons. For instance, instead of herbicide resistant rice, will a rice engineered genetically to be resistant to a serious pest be allowed for exchange unhindered? Are our objections to the transgenic per se or to the threats some transgenics pose to environment? For some, the transgenics are bad even if the alternative of extraordinary environmental damage by chemical pesticides keeps on taking place? In addition to the moral issues involved in hazards to soil microbial diversity, human health, and other diversity through pesticides, the issue also raises equity issues. Will poor people unable to afford chemical inputs be able to benefit from transgenics or not? And if they do, does moral dilemma become lesser on that account?

Case 2: Whether inserting human genes in animals/plants is ethical even if efficient to produce desirable compounds? A large number of people suffer and bear the cost of treatment for various ailments such as diabetes in which a single compound such as insulin can remedy the inadequacy. The insulin can be manufactured biochemically but also through plants and animals. In such a case eating a banana having an insulin expressing gene might provide enough daily dose to a patient. The cost may come down and a larger number of people may benefit. However, the issue is whether such a path to solve a problem is ethically sound and advisable. Let us look at some of these issues with the help of a case study provided by Comstock (1996b).

Years ago scientists "microinjected the pieces of DNA encoding the production of human somatotropin into the nucleus of fertilized pig eggs. The extracted embryos were reimplanted into the uteruses of sows, the pregnant animals came to term, and the first piglets in history with human genes were born" (Comstock, 1996b). The idea was not to produce bigger hogs but more effective pigs that could convert grains into lean meat faster than their parents while eating lesser grains than they did. This could be a boon to industry, consumers, and farmers. Among the nineteen transgenic swine, several expressed the higher level of growth gene but none achieved what was expected. Some suffered from medical problems, developed deformed bodies, and skulls. Others had swollen legs, ulcers, and renal diseases. Many were suffering from decreased immune function.

It is obvious that animals suffered as they do when we slaughter them for meat or for game. Does the suffering of nineteen pigs justify the goals, which if achieved, would benefit all the stake holders? How do we compare suffering of pigs with that of the farmers who may have to go out of business due to uneconomical production costs in certain conditions?

Many of these questions have been raised by Comstock (1996b) while dealing with the case of pigs with human genes. It is difficult to answer these issues one way or the other. The animal right activists oppose experiments on animals and a large number of pharmaceutical firms have done away with them. Instead, they use cell lines or other assays. Human cell lines have also been used for similar purpose of research and diagnostics. At what level of biological reductions do we bring ethical issues into the picture—at the level of DNA, cell, organ or body (living as well as dead)? Does responsibility lie in proportion to biological differentiation–the higher the differentiation, the greater the responsibility? Does it lie in respect of ability of a living being to envision future, bequeath memories, and traditions?

Brody (1995) in a forceful argument on patenting transgenic animals, makes a case for stronger regulatory framework that governs research on animals and the release of transgenic animals into the wild and justice among countries. He questions the moral case against the patenting of transgenic animals because (a) the classical Judeo-Christian tradition did underline the fact that "God created all things for man's sake"—a view which is obviously not shared widely now. The idea of human beings acting as steward or trustee of nature has also been questioned. Brody argues that the concept of trusteeship does not imply holding a property in trust unchanged. If a change in the property is necessary (past, present, or future) to serve the interests of its human owners, it should be done.

This notion is obviously at variance from the one held in many faiths which consider the rights of human beings not being superior to the rights of other living beings. And at the same time, many of these faiths have developed unique pragmatic solutions to deal with the ethical dilemma in day-to-day life. For instance, Buddhists in Bhutan and some other countries abhor violence and yet consume meat. The person who kills is considered of a lower status and social esteem but those who partake in this do not suffer from any guilt. While developing policies, the paradox becomes even more manifest. Shooting an animal which strays into private fields is acceptable, but if the animal, after being attacked, moves into the forest, the person can be charged for having perpetrated violence in the wild. Therefore, while Brody's argument about human concern for maximizing its long-term well being may be valid, in its essence, it also includes the option of future generations interpreting their possibilities of well-being afresh. However, if biological diversity and ecological integrity has been fundamentally altered, the options of future generations are compromised.

Cobb (1993) argues that maximizing value or well being for oneself for the rest of one's life might appear ethical but may fall if the concern for maximizing value for all human beings for indefinite future is brought into the picture. Here again the logic is utilitarian and in conformity with Judeo-Christian tradition. However, to increase well being of all living beings is certainly an ethical position well enunciated in Hindu philosophy and also in Buddhist's tradition. In some sense, the Kantian position of deriving ethical principles not on the basis of consequences but on their intrinsic interest or rightness is another extreme view. Cobb recalls Kantian argument for an ethical action to be generalizable and also applicable for various relationships in the past as well as the future. In reality, as Cobb rightly argues, it is impossible to take actions and be comfortable with them while ensuring one's consistency with all past and future commitments, obligations, and debts.

Human beings shape their criteria as well as the context in which they weigh the moral judgments progressively. The role of ethics is in ensuring a constancy of concern for certain core values. My submission is that this core value is conservation of nature (whatever little is left of it) in a manner that future options of the coming generations are not compromised irreversibly. The point to be stressed is that future generations will comprise the offspring of those who have options of surviving and procreating in the present. For millions of stillborn children and other disadvantaged human beings, there is no option of contributing votes in the future generations. Are we, therefore, only concerned with maximizing future options of those who fulfill the requirement of survival of the fittest?

This argument sometimes can be turned on its head to destroy nature. Many Third World activists argue that if the European cultures accumulated wealth by destroying the environment, why should the developing countries be denied the similar option? Such an argument has a ring of justice around it. However, it obviously implies that justice within the Third World can be achieved by destroying nature which is not only impossible but also unfair to a large number of communities and poor whose primary source of survival is whatever access they have to natural resources. The argument still remains that ethical responsibility of those whose life style may contribute far more but cannot be equal to those living on the edge. Ethically a poor person commits a wrong when he or she in one's short-term interest endangers the diversity irreversibly. Whether this violation is bigger than the violation by those who did not provide access to such people for alternative means of livelihood is an issue not adequately discussed in literature.

There are strong ethical traditions which justify people sacrificing their lives for upholding the rights of other living beings. There are tales of several Chola kings who established the principle of equal responsibility for human and non-human sentient beings. One of the kings was passing through a forest in his chariot. When he felt thirsty, he stopped near a spring. He went to the spring and came back after relaxing for a while. On his return he noticed a vine entwining around the spokes of the wheel of the chariot. The vine would have been broken if the wheel had moved. He thought for a while and left the chariot there in deference to the right of the vine and walked back to his palace (Honey Bee, 1994).

Another Chola king had a bell of justice hanging in a temple. The idea was that anyone feeling aggrieved could ring the bell and ask for justice from the king.

It was obligatory on the part of the king to hear the woes of the aggrieved party and dispense justice. One day the bell was ringing furiously. The king was disturbed and decided to walk to the temple himself to find out who was so aggrieved today. When he went there, he saw a strange sight. A lot of other people had also gathered in the meanwhile. There was a cow who was ringing the bell. On seeing the king she came down and started walking in a particular direction. The king also started following her. The cow stopped near a dead bullock—her son. The king immediately understood the situation after noticing the chariot with which the bull calf had had the accident. The chariot belonged to his son who had negligently run over the calf. As the story goes, he called for his son, asked him to lie on the ground, and asked that the chariot run over him (Honey Bee, 1993).

One may not agree with this notion of justice, and yet these are lampposts illuminating the path for a large number of people for whom the right of nonhuman sentient beings cannot be subordinated to the rights of the human beings. Cobb (1993) tries to reconcile the divergence of perspectives on the subject and suggests, "we now know that the others are not only human beings but, as the Hindus and Buddhists say, all sentient beings are, as the Old Testament says, 'all flesh'."

As you would see next, the normative position in different societies does often vary from the widely practiced position. It is the dilemma which common people face while choosing their point of reference in traditional versus contemporary morality that requires reflection in the current debate on bioethics.

Does each specie have a right to exist as itself? Don't mules, a cross between a horse and a donkey have a place in several cultures (Macer, 1994)? Macer questions whether species must necessarily be maintained only as they have evolved without any change whatsoever? What would then happen to the conventional animal and crop breeding, or for that matter what kind of modification of a genome of a specie is ethically acceptable and which will violate the sanctity of life (ibid)? Is producing insulin at a low cost by transferring genes for producing the same in a crop specie for wider human use acceptable or not? Inserting human genes into animals perhaps causes lesser ethical dilemma than vice versa. Macer recalls the argument of Singer which stipulates that human suffering is of greater importance than that of other animals simply because animals apparently do not have future plans while humans have. Is it the death of a future vision humans pursue which makes ethical concerns so central to human consciousness? If the feeling of pain extends the boundary of human responsibility towards animals, then will cessation of the feeling of pain through genetic manipulation reduce or eliminate ethical responsibility? Will the "human" treatment of animals make their rights dispensable because they cannot produce inheritable knowledge and, thus, cannot bequeath one to the future? The decisions for conservation must then be made by us in full recognition of this incapacity of nonhuman sentient beings. Much as we may argue and talk about the rights of other beings, Macer perhaps rightly asserts that it ultimately boils down to our duties (and responsibilities) towards other species. Whichever way they are defined.

The only problem with the above argument is that we may reduce the concept of responsibilities to merely utilitarian logic. In other words, if some species are not useful in the present reckoning, should their conservation be dispensed with? The conservative thinkers would suggest that resources being limited, tradeoff in allocation of resources is inevitable. The urgent, tangible, and human-oriented needs often take precedence over the long-term, not so tangible needs of non human beings.

So long as the cultural perceptions vary about the role of humankind in dealing with nature, their assessment of human duties towards nature will vary. Consequently, the risks that are perceived to be real, as different from imaginary, will be a function of respective discount rate about expected returns from various investments.

It is not these variations which often are the root of ethical dilemma. The problem arises when we consider certain positions decidedly superior to all others in various conditions. That eliminates possibilities of dialogue. So long as dialogue is on, it is possible to persuade and bring about change in other's positions. Biodiversity conservation requires placing ethical concerns for other non human beings at the center of our attention. But sometimes, we might achieve the results contrary to what we wish by sacrificing the concern for disadvantaged human societies for whom the future options arise only when they survive and procreate. However, we should also keep in mind that most decisions taken in the name of poor often never benefit them. Thus, we should not lose concern for other sentient beings and not achieve anything for the poor either. The tradeoff, thus, is often between interests of the poor and conservation of biodiversity. If such were the case, why would regions of high biodiversity also have high levels of poverty? The ethical values of the poor generally are stronger. It is the short-term interests of those who have no patience with nature and vainly believe that they can always create diversity in the laboratory to replace natural diversity that hurt the ecosystems most. Cutting corners once creates many times a cast of mind which propels cutting every corner without generating ethical dilemma. This, then, is the real danger when we do not even suffer from our suicidal follies.

Conclusion

In this paper I have covered three issues. One is what kind of dilemma one has to face while dealing with the development and diffusion of technologies requiring major transformation of ecological or biological integrity of species. I have argued that one cannot use only one set of principles while exploring the choices and options in different cultural and socioeconomic contexts. It will be futile to attempt imposition of one set of ethical values on all the cultures facing different developmental dilemma. At the same time, I have suggested that even in tradition most cultures have learned to evolve compromises that provide a practical way of dealing with the dilemma of coexistence. The second part of the paper has dealt with the ethical guidelines developed by Pew Conservation Scholars with the help of several background papers developed by the author and others. The discussion on the guidelines also includes the issues which these guidelines don't address. For example, the responsibility of the consumers and civil society in generating ethical conduct in accessing biodiversity could not be addressed in the draft guidelines. It is obvious that no amount of normative prescription will be able to check the unethical practice of dealing with biodiversity if the civil society does not feel disturbed and concerned about it. In the past, civil society mobilization did achieve major successes–for instance, in the case of advertising of powdered milk being at a par or superior to breast milk. However, similar consensus on the current issues of biodiversity is yet to be evolved despite the enactment of CBD and evolution of the earth charter. To that extent the task remains and this paper may help in extending the debate from a Third World perspective.✧

Selected Reading

Alcorn, B. J., Nov. 10–14, 1992. Conservation of cultural and biological diversity frames of discourse, analysis and action. Third International Congress Ethnobiology, Mexico City.

Axt, Lee and Ackerman, 1993. Biotechnology, indigenous peoples and intellectual property rights, p. 13. Congressional Research Service of US in its report of April 16, 1993.

Baenziger, S. P., Kleese, R. A., and Barnes, R. A. 1993 Intellectual property rights: protection of Plant Materials. Crop Science Society of America, Amercian Society for Horticultural Science, ASA, SSSA, USDA, CSSA. Special Publication Number 21, Wisconsin:CSSA.

Bozicevic, 1987. Distinguishing "products of nature" from products derived from nature, 69 J. Pat. & Trademark Off. Soc'y 415, 1987.

Brody, Baruch A., 1995. On patenting transgenic animals. The Ag Bioethics Forum, 7(2):1–7.

Chapman, R. A., 1993. Human rights implication of indigenous people intellectual property right, Agenda for an October 1993 conference of American Association for the Advancement of Science (AAAS).

Chisum D., 1991. Patents: a treatise on the law of patentability, validity and infringement, 1.02 and 1.03.

Cobb, 1993. Ecology, ethics, and theology, p. 211–228. In: Herman E. Daly and Kenneth N. Townsend (eds.), Valuing the earth. The MIT Press, London.

Comstock, 1996(a). Two case studies: pigs with human genes. The Ag Bioethics Forum 8(2)7.

Comstock, G., 1996(b). Commentary on herbicide resistant rice case study. The Ag Bioethics Forum 8(2)2.

Cunningham, A.B., 1993. Ethics, ethnobiological research, and biodiversity. World Wide Fund for Nature. Switzerland.

Dennis v. Pitner 106 F.2D 142—7th Cir. 1939 in CRS Report, 1993:49, p. 14.

Guidelines for country studies on biological diversity. United Nations Environment Programme, Kenya.

Gupta, 1991, 1992.

Hamilton, D. N., 1993. Who owns dinner: evolving legal mechanisms for ownership of plant genetic resources. Tulsa Law Journal Vol. 28:587.

Hamilton, D. N., Ellis, and Lellitt, N., 1994. Do we need legal and legislative remedies? American Association for the Advance of Science (AAAS), February 22, 1994, San Francisco.

Hecht and Cockborn, 1959. Forest peoples' manifesto: platform of the National Rubber Tappers Council (Appendix).

Hodges, T., Graveel, J., Joly, R., Vorst, J., 1996. Ten case studies innon medical bioethics. The Ag Bioethics Forum 8(2)1–2.

Honey Bee, Cover Story 1(1), 1994 (Newsletter of Farmers' Creativity and Innovation).

Honey Bee, Cover Story 4(2&3), 1993.

Hussan, A., 1991. Economic aspects of exploitation of medicinal plants, p. 125–140. In: O. Akerele, U. Heywood, and H. Synge (eds.), Conservation of medicinal plants. Cambridge University Press.

Khalil, H. M., Walter, R.V., and Juma, C., 1992. Biopolicy international: property rights, biotechnology and genetic resources. African Centre for Technology Studies, Kenya.

Macer, 1994:21.

Megarry V.C., 1977. In: Tito v Waddell (No.2) (1977) 3 all E.R. 129 at 222—a decision of the English Chancery Division Court.

Nietschman and Churcher, 1994, p. 2.

Nijar, G. S., 1994. Towards a legal framework for protecting biological diversity and community intellectual rights—a Third World perspective. Third World network discussion paper.

Reid, V. W., Laird, A. S., Gamez, R., Sittenfeld, A., Janzen, H.D., Gollin, A. M., and Juma, C., 1993. Chapter one: a new lease on life. WRI, Washington.

Shelton, Dinah. Fair play, fair pay: strengthening, local livelihood systems through compensation for access to and use of traditional knowledge and biological resources. Report prepared under the auspices of the World Wide Fund for Nature.

Sittenfeld, A. and Villers, R., 1993. Exploring and preserving biodiversity in the tropics: the Costa Rican case. Current Opinion in Biotechnology 4:280–285.

Snell's principles of Equity, 28th edition, 1992, p. 192. Beatty v Guggenhelm Exploration Co (1919,) 225 N.Y. 380 at 386 (Cardozo J.).

WHO, IUCN and WWF, 1993. Guidelines on the conservation of medicinal plants. The International Union for Conservation of Nature and Natural Resources (IUCN), Switzerland.

Yamin, F., Posey, D. Indigenous peoples, biotechnology and intellectual property rights. Indigenous Peoples 2, Number 2.

Ecological Context of Biodiversity

K. G. Saxena,[1] K. S. Rao[1] and P. S. Ramakrishnan[2]
[1]Sustainable Development of Rural Ecosystems Programme, G.B. Pant Institute of Himalayan Environment and Development, Kosi-Katarmal, Almora, India;
[2]School of Environmental Sciences, Jawaharlal Nehru University, New Delhi, India

Abstract

Species, the fundamental units of observation in ecological assessment of biodiversity, are unevenly distributed and occur in an intricately spatial mosaic, classified on a global scale into biogeographic zones, biomes, ecoregions, oceanic realms, centers of endemism and centers of origin and domestication (in the context of agricultural biodiversity). Considering the highest level of species richness in the tropical rain forests, hot and humid climate in the tropics could be viewed as a primary environmental indicator of biodiversity.

The environmental controls on biodiversity are, however, intricately linked with geological and evolutionary processes in the distant past, and human factors related to population pressure, technology, market, and sociocultural changes in the recent past. Within global regions, biodiversity has been found to be related to a variety of factors such as geographical isolation, environmental heterogeneity, landscape processes and nature, frequency, and intensity of natural/man induced disturbance.

All ecosystem services are affected to one degree or the other by the changes in biodiversity. Ecosystem properties, such as resource use efficiency, productivity, and stability, are influenced by biodiversity. Diverse ecosystems, natural or manmade, are generally more efficient in capturing energy, water, nutrients, and sediments than the homogeneous systems. Biological diversity is concentrated in areas inhabited by socioeconomically marginal traditional societies. Biotic complexity is a key indicator of sustainability and buffering capacity. Ecological context of biodiversity is reviewed in this article.

Table 1. Density of medicinal plants in major natural vegetation types in Chhakinal watershed

Species	Pine forest (1600–1750)	Cedrus forest (1750–2000)	Abies forest (2000–2300)	Aesculus forest (2300–2900)	Quercus forest (2600–2900)	Betula forest (2900–3200)	Alpine meadow (3200–3500)
Anemone obtusiloba D. Don.	—	—	—	—	—	—	10400
Juglans regia L.	—	—	12	53	—	—	—
Leucas lanata Benth.	—	—	—	—	8000	—	—
Picrorrhiza kurrooa Royle & Benth.	—	—	—	—	48300	8000	24000
Plantago major L.	—	—	—	34400	61200	2400	2400
Polygonum amplexicaule D. Don.	—	—	93200	45200	58400	74400	34400
Ranunculus laetus Wall.	—	—	—	—	10400	—	18400
Rhododendron campanulatum D. Don.	—	—	—	—	1730	11	413
R. anthopogon D. Don.	—	—	—	—	—	—	1293
Taraxacum officinale W. Wigg.	—	—	—	—	—	—	5200
Taxus baccata L.	—	—	46	—	—	—	—
Number of medicinal plant species in the community	0	0	3	3	6	4	8
Number of other species in the community	21	23	33	40	37	27	34

(Source: Dobariyal *et al.*, 1996)

Introduction

Biological diversity has a variety of contexts (Heywood and Watson, 1995; Ramakrishnan et al., 1996). Biodiversity is dealt at three distinct levels: genetic (within species), species (species numbers), and ecological (community) diversity. Ecological perspective of biodiversity covers interactions between different constituents of the community in the context of the physical environment. Spatial-temporal variations in biodiversity, factors regulating biodiversity and biodiversity-ecosystem stability relationships have been important ecological concerns for quite a long period.

Biodiversity is not only an important determinant of ecosystem processes and responses, but is also crucial for satisfying a variety of human needs. Inclusion of human beings in ecosystems leads one to relate biodiversity with sustainable development concerns (Lovejoy, 1995).

Geographical Patterns of Biodiversity

Latitudinal and Attitudinal Gradients

Attempts on identifying geographical trends of biodiversity distribution have considered species richness as the currency of diversity. Tree species richness has been shown to increase with decrease in latitude (Latham and Ricklefs, 1993). The effect of latitude on species richness varies from one group of plants to the other and from one region to the other. Tree species diversity declines much more steeply from equatorial regions into southern temperate zones than it does to the north (Currie, 1991) indicating variability in latitude effect in different geographical regions.

Temperate forests of east central Asia, eastern North America, and Europe have an overall tree species richness ratio of 6:2:1 though the three regions have similar latitudinal spread. The number of plant species recorded in African continent (low latitude region) is 10 times of that in the Europe (high latitude region) but there is nominal difference in the number of polypore fungi recorded in the two continents (Ryvarden, 1993). Ricklefs (1977) did not find any difference between temperate prairies of central North America and tropical grasslands in terms of species richness.

Varied patterns of diversity are reported along attitudinal gradients. Gentry (1988) reported decrease in woody species richness with increase in altitude.

Lieberman et al. (1996), based on intensive sampling across an attitudinal gradient from 30 m to 2600 m in Costa Rica, reported highest species diversity a 300 m and progressive decrease in species richness both above and below this altitude. Such a humped pattern was also reported by Arroyo et al., (1988) across aridity gradient in Chile Andes. Community analysis across an elevational gradient of 1600 m to 3600 m in a Himalayan watershed showed increasing species richness trend up to 2900 m, followed by a decline in 2900–3200 zones and increase in 3200–3500 m zone (Table 1).

Many economically important families, such as Annonaceae (custard-apple family), Lauraceae (cinnamon family), Moraceae (fig family), Dipterocarpaceae (Dipterocarp family), Ebenaceae (ebony family), and Meliaceae (mahogany family) are almost entirely restricted to the tropics. Within the tropics, neotropical region as a whole is much more species diverse than the palaeotropics (Table 2). Within neotropics, the northern Andean region is rich in herbaceous species, whereas Amazonia is rich in trees and shrub species (Prance, 1995).

Table 2. Comparison of plant taxic diversity of three major tropical regions

	Africa	Malesia	Neotropics
Families	271	310	292
Genera	3750	3250	4200
Species	40000–45000	42000	90000

(Source: Prance, 1995).

Latitude or altitude *per se* do not drive biodiversity gradients. Factors that also vary with latitude and altitude are climate and energy flux. The latter are correlated with evapotranspiration and ecosystem productivity. Compilation of global data shows a positive relationship between tree species richness and actual annual evapotranspiration. While a trend of increasing diversity towards the tropics could be generalized, prediction of species richness from the environmental factors or geographical attributes may not be precise.

Centers of Diversity

Centers of diversity are the areas of exceptionally high levels of diversity or those having unique assemblage of species such as endemism. Diversity and endemism are produced by different processes. Endemism is the product of isolation.

Diversity is achieved through isolation, allowing genetic differentiation between populations and exchange of populations between semiisolated areas allowing newly allopatrically formed species into sympatry. Endemic species richness may or may not be correlated with phylogenetic diversity or total species richness. Mediterranean flora is an example of richness of total species as well as endemics (Cowling et al., 1992; Arroyo et al., 1995). Floristic patterns in India indicate that endemics, as well as total species of the country, are concentrated in the Western Ghats and the Himalaya (Rao, 1996).

Table 3. Biodiversity hotspots

Hot spots	Habitat type
Atlantic coast of Brazil	Tropical rain forest
Colombian Choco	
Northern Borneo	
Peninsular Malaysia	
Philippines	
Western Amazonia Uplands	
Eastern Area Forests of Tanzania	Tropical montane forest
Western Ghats in India	
Western Ecuador	Mixture of tropical rain forest and tropical montane forest
Madagascar	Tropical moist forests
Cote d'Ivorie	
Sri Lanka	
Eastern Himalaya	Subtropical and warm temperate forests
California	Mediterranean vegetation
Cape Floristic Provinces	
Central Chile	
South Western Australia	

(Source: Myers, 1990)

Centers of diversity could be recognized at the global, regional and local scale. The centers would contain more species or more of some other measure of biodiversity than would be expected from their environmental predictor variables. Myers (1988) recognized ten pockets constituting 3.5% of the remaining forests that harbor 27% of all tropical forest plant species. A more expanded analysis led identification of 18 "hot spots" supporting 20% of world flora on only 0.5% of area of the earth (Myers, 1990). These hot spots include tropical rain forest, tropical montane forest, tropical moist forest, subtropical-warm temperate forests, and Mediterranean vegetation (Table 3). International Union of Conservation of Nature and Natural Resources has identified 234 areas of specially high plant species diversity. In addition to species richness and endemics, evolutionary peculiarities, habitat diversity, and threats to conservation were also considered in the identification of areas (Table 4).

Table 4. Criteria adopted by IUCN for identification of centers of diversity

✓	The area is known is evidently rich in species, even though the number of species present may not be accurately known.
✓	The area is known to contain a large number of endemics.
✓	The site contains an important gene pool of plants that are of value to humans or that are potentially useful.
✓	The site contains a diverse range of habitat types.
✓	The site contains a significant proportion of species adapted to special edaphic conditions.
✓	The site is threatened or under imminent threat of large-scale devastation.

Some regional centers of diversity contain large number of closely related genera; e.g., fynbos habitat of Cape Floristic Region has 526 species of *Erica* largely due to isolation and proliferation of non-fire-tolerant traits (Cowling and Holmes, 1992) and Australia has 600 species of *Eucalyptus* as a result of creation and rejoining of forest refugia (Groves, 1994). On a local scale, centers of diversity would be the biodiversity rich areas within a given country which may or may not have regional or global importance.

Factor of Spatial Scale in Diversity Assessments

Whittaker (1972) characterized diversity at different scales: alpha, beta, and gamma components. Alpha diversity is measured as the number of species in potentially interactive assemblage of species. This measure of diversity would be most valid when sampling is done in ecosystems at the finest spatial scale. Gamma diversity is the overall diversity within a large region, usually to a country or region but rarely to a continent. Beta diversity is a measure of degree of species change along a habitat or physiographic gradient.

Within a given ecosystem type identified at a local scale, species richness could increase with increase in area or individuals censused. Tree species richness generally reaches an asymptote at 1–3 ha (Gentry, 1988; Toumisto et al., 1995). Contrary to this generalization, Condit et al. (1996) showed continued accumulation of tree species up to and beyond 50 ha (Figure 1). These results showed that species-individual curves could be more useful for comparing and assessing diversity than species-area curves.

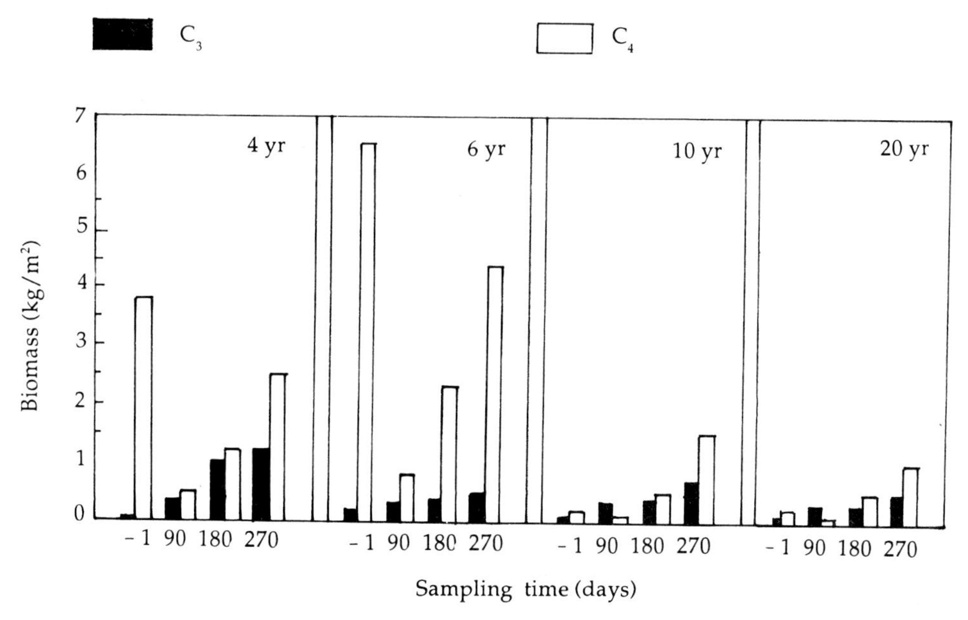

Figure 1. Species area curves from three 50-ha plots, three different dbh classes at each, plotted on log scales (Source: Condit et al., 1996).

The impact of the factor of spatial scale has been formerly explained from species richness in islands varying in size (MacArthur and Wilson, 1967). The reasons attributed to an increase in species richness with the increase in island size are (a) populations on large islands are large enough to make extinctions less likely and (b) larger islands contain the specialist species of a greater number of habitat types. Kohn and Walsh (1994) attempted to segregate the effects of habitat

Table 5. Summary of thc most frequently cited mechanisms by which plant species coexistence is promoted in natural communities*

Theory	General Description	Principal Assumptions	Reference
Exploitative competition—			
Resource partitioning	Species differ sufficiently in their ability to exploit diffierent limiting resources such that competitive exclusion does not occur	Tradeoffs exist in species' abilities to compete for two or more limiting resources	Tilman (1982)
Spatial heterogeneity	The environment is composed of numberous distinct microhabitats in each of which one species competitively excludes all others	There are at least as many different microhabitats as plant species	Pacala & Roughgarden (1982)
Neighborhood effects	Intraspecific spatial aggregation of species serves to increase the relative strength of intra- rather than interspecific competition and thus facilitates coexistence	Plant spatial distributions are typically intraspecifically clumped and some times interspecifically segregated	Pacala (1986)
Temporal heterogeneity	Temporal variations in "favourable" periods ensure that different species will be the dominant competitions at different times	Communities are composed of long lived and highly fecund fugitive species	Warner & Chesson (1986)
Preemptive competition—			
Regeneration Niches	Species differ sufficiently in their regeneration requirements that competitive exclusion does not occur	The environment is composed of numerous distinct microsites	Grubb (1977)
Disturbance	Species richness should be greatest at intermediate levels of disturbance since dominance is prevented and the pool of potential colonists is relative large	Habitat productivity is inversely proportional to the degree of disturbance	Grime (1973)
Lotter models	Seeds colonize vacant microsites at random, resultant adults are competitively superior for the microsite and sufficiently long lived and fecund to ensure further colonisation	Species are long lived and highly fecund, probability of colonisation is a simple function of seed production	Sale (1977)
Patch-dynamics	The dispersal ability of the inferior competitor is sufficiently great for its rate of successful colonisation to be higher than its rate of extinction from microsites	Tradeoffs exist between competitive ability and dispersal	Levis (1974)

** primarily associated with mediating exploitative competition and primarily associated with mediating preemptive competition (Source: Hulmes, 1996).*

diversity and size of the island on species richness by measuring richness in random quadrats (small scale diversity), as well as for the entire island (large scale diversity). An increase in small-scale species richness with an increase in the island size was explained as the area effect and that of the whole island due to area effect as well as habitat diversity effect. The two effects were equal in magnitude.

An understanding of the relationship between local and regional diversity is extremely important for conservation planning and management. This is an area which needs more attention (Franklin, 1993).

Ecological Theories of Species Diversity

Species diversity is a product of histories of species' accumulation and disappearance. A species is gained by migration from outside or from production of new species within the regions. The species' loss could be due to extinction, unpredictable drastic changes in physical conditions, stochastic events, and biological factors, including competition, predation, pathogens, and dispersal agents. A number of hypothesis have been conceived to explain biodiversity patterns (Table 5). There is still no consensus regarding the mechanisms underlying species diversity (Crawley, 1986; Tilman and Pacala, 1993).

Equilibrium Versus Nonequilibrium Theories

Equilibrium theory suggests that habitats become saturated with species and the saturation limit is determined by the outcomes of local interactions of species. The system tends to compensate for species losses due to extinction through gains as a result of speciation. Nonequilibrium theories suggest that communities never reach a state of equilibrium of diversity, and change in species richness at a given point of time or place is determined by the effect of environment on species production, exchange, and extinction processes. Equilibrium theory would assume that rare species would always be rare. The nonequilibrium theory would assume that rarity or abundance could be transient unstable phase. The two theorics are alternative explanations. Wiens (1984) pictured communities as existing along a spectrum from stable equilibrium systems to nonequilibrium systems. The former systems are structured primarily by biotic factors (self-regulating population growth, compensatory interactions) while the latter are controlled more by environmental variations (strong stochastic effects like storms, temperature changes)

(DeAngelis and Waterhouse, 1987). If vast areas are identified as a single ecosystem, it could represent a mixture of microsites under equilibrium and nonequilibrium states.

Spatial-temporal Heterogeneity and Niche Segregation

Niche segregation could be related to evolution of diverse patterns of growth, reproductive strategies, photosynthetic pathways, and nutritional relationships. Stable coexistence of trees and grasses in the Savanna ecosystem is attributed to exploitation of deep soils by trees and surface soil by the grasses under water stress condition (Walker, 1985; Belsky, 1990). Coexistence of grass and legume constituents benefit each other in grassland ecosystems such that mixture of the two use resources more effectively than their corresponding pure strands. However, the advantages of mixture are observed in grass-legume combinations of historical coexistence (Turkington and Jolliffe, 1996).

Coexistence of a diverse early successional weed community following fire under shifting agriculture could be due to complementary patterns of growth, reproductive strategies, and nutritional relationships associated with C_3/C_4 photosynthetic pathways and nonsprouting/sprouting strategies. The C_4 species grew faster during the early part of the growing season whereas C_3 grew faster during the later part of the growing season (Figure 2). The former were more tolerant towards nutrient stress compared to the latter. Nonsprouting species, coming up through small seeds, showed higher rates of growth and nutrient uptake than the sprouting species regenerating through underground vegetative organs. Sprouting species could compensate the disadvantages of slow growth and nutrient uptake rates through the strategy of accumulation of resources in the belowground organs and the transfer of these resources to support aboveground growth following fire (Saxena and Ramakrishnan, 1983).

Explanation of species diversity due to spatial heterogeneity is based on the assumption that there are at least as many different microhabitats as plant species (Pacala and Roughgarden, 1982) and that due to temporal heterogeneity on the assumption that communities are composed of long lived and highly fecund fugitive species (Warner and Chesson, 1985). Since tradeoffs exist in species' abilities to compete for different resources, species differ sufficiently in their ability to tolerate different limiting resources, such that competitive exclusion does not occur (Tilman, 1982).

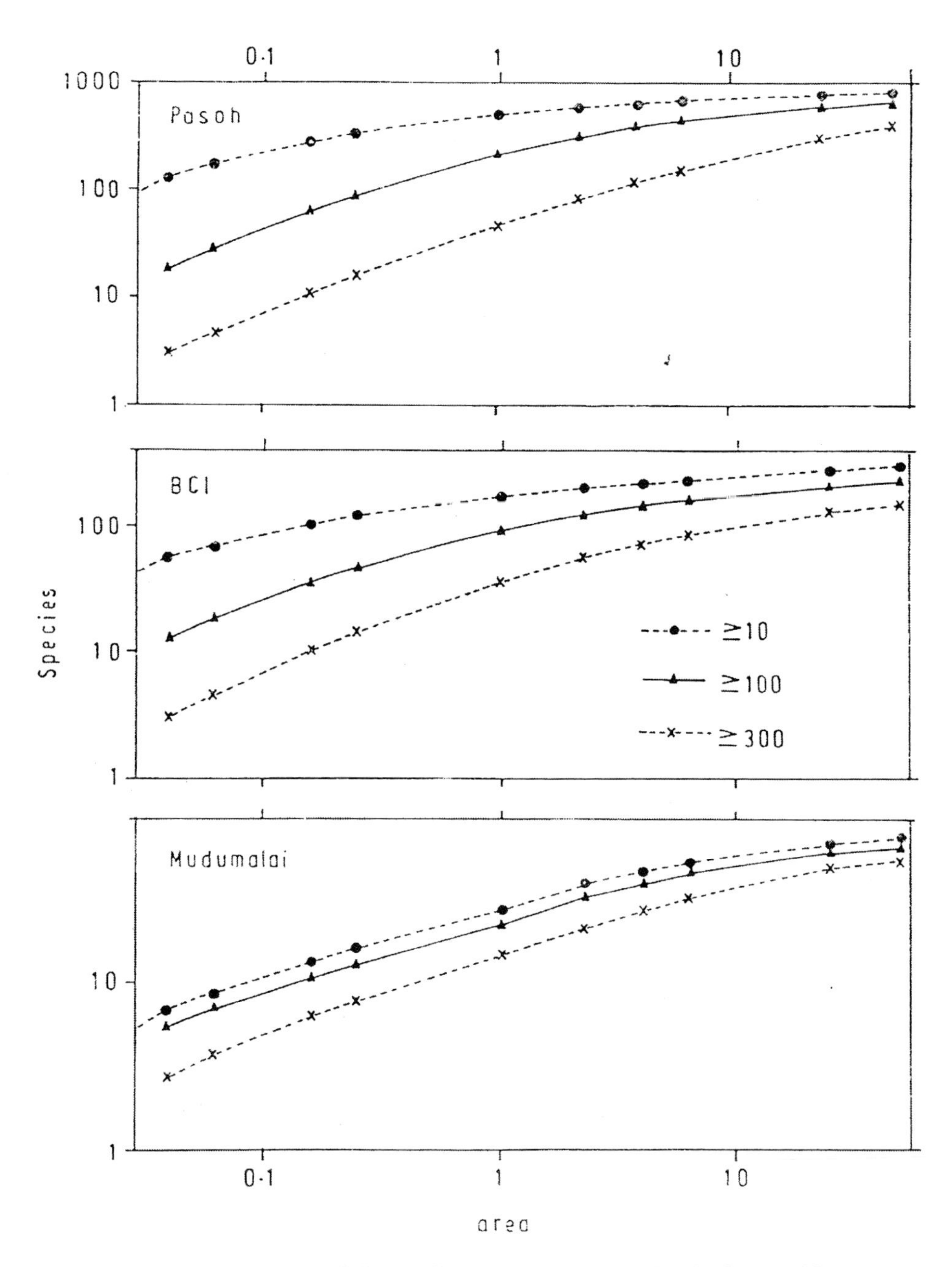

Figure 2. Changes in biomass of C_3 and C_4 species on plots under 4-, 6-, and 20-year jhum cycles, a day before the burn after slashing (-1) and at different times (days) after the burn (Source: Saxena and Ramakrishnan, 1984).

Neighborhood Effect Theory

Pacala (1986) postulated that plant spatial distributions are typically intraspecifically clumped and sometimes interspecifically segregated. Intraspecific aggregation is likely to result in a higher intensity of intra rather than interspecific competition, enabling species coexistence.

Neutral or Lottery Models

Neutral or lottery models rule out any significant role of niche segregation or competitive coexistence in supporting diversity. Fugitive species adapted to large-scale production of long dispersal mechanisms could survive with superior competitors by continually colonizing unoccupied space (Tilman, 1994).

Table 6. Floristic similarly based on the coefficient of community index of preburn herbaceous vegetation with that of uncropped and cropped sites under different Jhum cycles

Jhum Cycles (years)	Treatments	
	Preburn and uncropped sites	Preburn and cropped sites
4	98.0	74.1
6	66.7	70.6
10	14.8	20.0
20	14.3	20.0

(Source: Ramakrishnan, 1992)

Community Drift Model

This model assumes that a large number of competitively identical species with limited dispersal range change in abundance due to random drift (Hubbel and Foster, 1986). Rarity and narrow ranges guarantee that a species-accumulation curve would continue to increase at all scales until the boundary of continent is reached. An asymptote; i.e., a ceiling value for the upper limit of local species richness, is more likely when competition and other interactions determine community organization (Cornell, 1993). Neutral and drift models have been supported by coexistence of hundreds of tree species in a plot of one hectare (Gentry, 1988; Valencia et al., 1994; Condit et al., 1996).

High Probability of Predation Near Conspecific Adults

Janzen (1970) hypothesized that seeds are more likely to escape predation in the vicinity of adult conspecifics because the activities of seed herbivores tend to be concentrated here. This could result in a possible bias in favor of rare species (Clark and Clark, 1984).

Gap Dynamics

Gap dynamics has been found as an important mechanism supporting high species richness in tropical rain forests. Rain forest ecosystem considered as uniform community consists of a matrix of small gaps differing in respect of plant growth in an ecological time scale. Gaps develop as a result of natural death of old trees which could be influenced by stochastic environmental processes; i.e., periodic environmental fluctuations. The gaps could be as small as 30 m^2. Gap turn over rate (mean time between formation of a gap at any one point in the forest area) could vary between 60 years to 159 years. Species differ in their response to gaps (Chandrashekara and Ramakrishnan, 1994). Species able to germinate, establish, and grow under canopy and those coming up only after gap creation represent the two extremes (Swaine and Whitmore, 1988).

Gap phase dynamics could be viewed as an extension of a patch dynamics model proposed by Levin (1974) based on an assumption that tradeoffs exist between competitive ability and dispersal capacity. The high dispersal ability compensates for the risks of extinction from microsites as a result of inferior competitive ability.

Gap or patch dynamics theory has some analogy with explanation provided by Grime (1973) for disturbance as a factor influencing community structure. Disturbance prevents dominance and species richness is likely to be greatest at intermediate level of disturbance regime. There are several studies suggesting that disturbances like herbivory and fire are related species richness (Hulme, 1996; Jeltsch et al., 1996).

Landscape Ecology Approach

The conventional approach of looking ecosystems as self-contained ecological units is increasingly elaborated where different ecosystems are considered to be interconnected spatial units in a landscape. Extinctions at small spatial scales could be inescapable but may be prevented at large scale. Variability between different ecosystem types and their spatial configurations and interconnections together would determine the extinction, migration, and speciation in a landscape. It is more likely that an ecosystem type represents a transient rather than equilibrium state (Shugart, 1984). Because of variation in topography, climate, disturbance regime, and resource use history, landscape could be highly complex. Approaches that can integrate the transient patches into larger landscapes and can predict the spatial-temporal patterns of ecological communities over the landscapes are evolving (Heywood and Watson, 1995).

Table 7. Crop mixture and sequential harvesting of crops under shifting agriculture at lower elevation in Meghalaya in Northeast India

Species	Harvesting time
Setaria italica	mid-July
Zea mays	mid-July
Oryza satriva	early September
Lagenaria spp.	early September
Cucumis sativa	early September
Zingiber officinalis	early October
Sesamum indicum	early October
Phaseolus mungo	early October
Cucurbita spp.	early November
Manihot esculenta	early November
Colocasia antiquorum	early November
Hibiscus sabdariffa	early December
Ricinus communis	(perennial crop)

Note—All the seeds were sown in April.
(Source: Ramakrishnan, 1992).

Biodiversity and Ecosystem Function

The changes in biodiversity accompany changes in ecosystem functions. Several hypotheses have been proposed to explain the observed or expected consequences of species addition or removal.

Keystone Species

Paine (1969) opined that only one or few species, by virtue of their unit attributes, have uniquely important effects on community structure and function. The impacts of biodiversity change on ecosystems would be dramatic when abundance of these species is affected. Paine (1969) developed the concept of keystone species based on the observations of predators of competitive dominants. Elaborating this concept, Power and Mills (1995) referred keystone species as those whose impacts on community organization or ecosystem function are disproportionately large relative to their abundance. Exotic nitrogen fixing species in the Hawaiian islands (Vitousek et al., 1987), nitrogen-fixing indigenous species such as Nepalese alder, and potassium accumulating bamboo species in shifting agriculture land use system in northeastern India (Ramakrishnan, 1992) are identified as keystone species. The keystone role of a given species may be restricted to specific ecological conditions. Thus, nitrogen fixers or potassium accumulators would assume a keystone role in situations where nitrogen or potassium availability is low and not in the conditions where these nutrients are not limiting ecosystem productivity.

Redundancy/Functional Compensation Hypothesis

In contrast to the keystone species concept, Walker (1992) hypothesized that species overlap with one another in functional properties such that loss of any one species has negligible effect on ecosystem properties. This concept of evolution of the same functional attributes in a number of species has been termed as redundancy hypothesis (Walker, 1992) and functional compensation hypothesis (Menge et al., 1994).

The rivet hypothesis of Ehrlich and Ehrlich (1981) is a mix of the keystone species and redundancy concepts. It suggests that each species is like a rivet in an aircraft. Each species contributes to ecosystem function in a unique manner and there could be a few analogous to critical rivets in aircraft which may have more critical roles than the others.

Human Use of Biodiversity

From the human use perspective, four distinct types of values are attached to biodiversity-direct, indirect, optional, and intrinsic values. Biodiversity (looking at the genetic species or ecosystem level) is concentrated in areas inhabited by the traditional societies. These societies were presumed to be the major threats to sustained use of biodiversity. This assumption led to conventional management of the allocation of land-agricultural area (including horticulture and plantation forestry) for direct uses of biodiversity to meet the immediate needs of humans and protected areas for indirect, optional, and intrinsic values of biodiversity. The more recent concept of biosphere reserve is an articulation to reconcile the two extreme hypothetical views: 1) local people and their resource use practices threaten the conservation objectives and so should be kept away from the conservation area (this view holds true for the core zone), and 2) local people have a major role to play in achieving the objectives of nature conservation as they have a rich empirical knowledge of sustainable resource uses. The present richness of biodiversity is largely due to conservation attitudes of the local communities. This view holds true for the buffer zone. Knowledge on biodiversity perceptions of traditional societies and their implications for biodiversity conservation and management needs is limited. Some aspects on biodiversity management by traditional societies in the Himalaya, a mountain system distinguished for its biodiversity and ecological fragility, are discussed in this section.

Biodiversity Knowledge and Values in Traditional Societies

Sustainability in the traditional societies meant the goal of meeting the essential survival needs locally. Isolation and inaccessibility fostered identification of species directly useful to humans and their sustainable use (Farooquee and Saxena, 1996). The social systems valued a biocommodity in terms of the constraints a given family or group of families faced in its procurement in conjunction with ecological opportunities and constraints in production/regeneration. Monetary valuation of a commodity, a concern originated from nontraditional urban/industrial/commercial society, is a recent change induced by the aliens in the traditional value system. Traditional societies tuned to subsistence economy for generations failed to integrate themselves in the markets. This failure could be attributed to a number of factors including population pressure, structural reforms in traditional land tenure and resource endowment systems, poor integration of technological advancements in traditional knowledge, and trade policies unfavorable to the local communities.

Traditional livelihood systems are adapted to a diverse plant and animal products in contrast to modern systems relying on a select few species. This, in terms of ecological management, implies an efficient utilization of environmental resources and processes to the benefit of mankind.

Table 8. Area under different cropping patterns as percentage of total valley and terrace cultivation area

Cropping patters	Area (%)	
	Valley land	Terraced slope
Wheat (HYV)-paddy (IC)	99.5	—
Wheat (HYV)-paddy (RC)	—	0.4
Wheat (HYV)-mmaize	—	39.6
Wheat (HYV)-amaranths	—	7.2
Wheat (HYV)-fingermillet	—	2.8
Wheat (HYV)-potato	0.5	5.8
Wheat (HYV)-mixed crops	—	13.6
Wheat (LC)-mixed crops	—	3.4
Barley-maize	—	9.4
Mustard-paddy (RC)	—	0.3
Mustard-maize	—	17.7

IC, local irrigated cultivar; RC, local rainfed cultivar.
(Source: Singh et al., 1996).

Landscape Complexity

Complexity of landscape results from coexistence of a variety of land uses differing in respect of man-induced changes in natural patterns and processes. Traditionally, family rights were sanctioned for agricultural land and community rights for utilization of natural vegetation. The area under cultivation used to be kept to a minimum through informal community control mechanisms. Agricultural land in the Himalaya is in the form of patches in the matrix of forests. Agricultural productivity is sustained with energy and nutrient inputs derived from forests. Sustainable utilization of forests and sustainable agriculture are, thus, interdependent and complementary objectives. The landscape configuration and linkages could differ from one region to the other depending upon the sociocultural and ecological factors (Ramakrishnan et al., 1996).

Agribiodiversity

In the northeast where shifting agriculture is the predominant agricultural land use, upland landscape is a mosaic of patches representing current year slash-burn-cropped sites—sites representing second or third year of cropping following the burn, secondary successional sites representing different ages of fallow development, and the pristine ecosystems in "sacred groves." The contribution of a given ecosystem type to the overall biodiversity in the landscape could vary. An area subjected to short cultivation cycles of 4–6 years would be less diverse than an area under longer cycles of 10–20 years because species composition of 4–6-year-old fallow fields radically differs from that in 10–20-year-old sites (Table 6).

Mixed cropping under shifting agriculture (Table 7) could be viewed as a mechanism of generating environmental heterogeneity, fuller utilization of the environmental resources to meet the diverse requirements of the humans and of reducing runoff and soil loss. The number of crop species in a plot of 1–2 ha could be as high as 40 sown together and harvested at different times during the year. Shifting cultivation remains a sustainable land use as long as sufficient land is available to practice long cultivation cycles but accompanies loss of biodiversity, soil and water under the situations of longer cycles. The shortening of the cultivation cycle is partly due to increase in population pressure and is also due to radical changes in forest land ownership rights (Ramakrishnan, 1992).

The landscape in the central and northwestern Himalaya is characterized by settled upland terraced agroecosystems. Heterogeneity in the agricultural land is created through rotation of crops in time and space. Mixed cropping in these areas does exist but to a much less extent as compared to shifting agriculture in the northeast (Table 8). Use of farmyard manure, land preparation, and draught energy input further distinguish settled upland agroecosystems from the shifting agriculture in the northeast. Farmyard manure consists of leaf litter collected from the forest floor and excreta of the livestock which derive a substantial portion of their feed from the forests.

Table 9. Number of crop varieties grown in organic agroecosystems*

Crops	Number of crop varieties		
	Chowda	Khaljhuni	Mikhila
Amaranthus paniculatus	0	2	2
Eleusine coracana	1	1	1
Fagopyrum esculentus	0	1	1
F. tataricum	0	1	1
Hordeum vulgare	1	2	2
H. winlense	0	2	2
Oryza sativa	3	0	0
Phaseolus lunetus	0	4	4
Solanum capsicum	2	1	1
S. tuberosum	1	1	1
Triticum aestivum	2	2	2

**Villages Khaljhuni and Mikhila, and inorganic fertilizer dependent agroecosystems of village Chowda in Kapkot Block in Central Himalaya (Source: Rao and Saxena, 1996)*

To sustain agricultural productivity in these organic agroecosystems, conservation and sustainable utilization of forest resources becomes a necessity. Reduction in crop biodiversity is often a result of forest destruction when farmers switch over from organic manure to inorganic fertilizer (Table 9). Because inorganic fertilizers are cash market commodities, their use becomes a catalytic factor forcing farmers to abandon traditional diversified crop systems that favor food security at the household level and to adopt cash crops and market dependent food security mechanisms.

Table 10. Minor forest products/nontimber forest products used in high altitude villages of Kuamon Himalaya (medicinal plants)*

Species	Uses	% families using the product
Aconitum heterophyllum	Decoction of tubers is used in intermittent fevers.	96
Aesculus indica	Seeds are used as veterinary medicine.	6
Angelica glauca	Seeds are used as corminative for digestion and delicacy.	12
Betula utilis	Decoction of bark is used as antiseptic lotion.	12
Fraxinus micrantha	Bark is used for bandage for humans and animals in fractures.	48
Juglans regia	Decoction of bark is used as antiseptic lotion.	60
Mentha arvensis/longifolia	Decoction used for stomach disorders.	12
Nardostachys jatamansi	Decoction from tubers and roots is used as antispasmodic drug; i.e., for palpitation of the heart and flatulence.	36
Ophelia (Swertia) chiretta	Plant parts are boiled and extractent is used as tonic for indigestion and also in malaria fever.	36
Orchis latifolia	Decoction is used as expectorant.	75
Picrorhiza kurrooa	Powdered roots are taken for stomach disorders and asthma.	96
Rheum emodi	Decoction from underground parts is used as medicine for diarrhea, jaundice and other stomach disorders; also used for muscular pains.	92
Rhododendron companulatum	Decoction of leaves used for colds and paste used for external application for rheumatism.	25
Thalictrum foliolosum	Decoction used as medicine for fevers and as health tonic after long illness.	12

* *(Based on Rao and Saxena, 1996)*

Forest Biodiversity

Timber exploitation, a resource use that accompanies drastic changes in the forest ecosystem and could lead to enormous soil loss and hydrological imbalance, has been restricted to the essential needs of survival. Even in shifting agriculture, the large trees are not slashed.

Trees coming up as a part of natural regeneration following the burn are not treated as "weeds." The long-term advantages of tree species regeneration in terms of soil conservation and the recovery of soil fertility are traded off with their short-term negative impacts in terms of lowering of crop yields due to the shade effect.

A few Naga tribes even plant keystone species like *Alnus nepalensis* to accelerate the fertility recovery process. Weeding of herbaceous species is also partially done as traditional farmers realize the positive role of these species in terms of soil and nutrient conservation (Ramakrishnan, 1992). It is also the case in the central and northwestern Himalaya with traditional societies that conserve many naturally regenerating tree species in settled upland terraced agroecosystems and recycle weed biomass through farmyard manure.

Traditional forest management concentrates on nontimber forest products (NTFP) in contrast to focus on timber in the conventional forest management. The species richness of NTFPs is much higher than that of timber species. The NTFPs include a range of plant groups and uses (Tables 10 and 11).

Sustainable use in traditional knowledge implied consideration of growth of individuals and the distribution of species. Commercial extraction of medicinal plant products from the village forests (forests owned and managed by the village community) only once in 12 years suggest people's perception of sustainable utilization differing from that in conventional forest management that permits utilization of the resource usually once in 4–6 years.

Obtaining timber, NTFP, and environmental conservation are recognized as alternative choices of management in the government forests, whereas the three forest uses are considered in an integrated perspective in the traditional management (Rao and Saxena, 1996). This mismatch between the traditional and conventional forest management leads to people-policy conflicts and engenders negative attitudes in local communities over the issue of protection of biodiversity in government forest lands. Obviously, threats to biodiversity loss are more likely in the forest land owned by the government than in the forests owned by the local communities.

Table 11. Minor forest products/nontimber forest products used in high altitude villages of Kuamon Himalaya (nonmedicinal plants)*

Species	Uses	% families using product
Wild edibles/food values		
Aesculus indica	Seeds used as staple in event of food shortage.	?
Ferns (3 species)	Used as green vegetables.	30
Fragaria chiloensis	Juicy fruits are eaten.	45
Juglans regia	Nuts are eaten as dry fruit.	100
Litseolea consimilis	Edible oil is extracted from seeds.	4
Morchella sps.	Highly nutritious vegetable.	15
Princepia utilis	Edible oil is extracted from seeds; fruits also eaten.	7
Prunus armenica	Pulp of fruit is edible. Oil is extracted from seeds.	60
Rubus niveus	Juicy fruits are eaten.	45
Urtica sps.	Used as green vegetable.	15
Fibres		
Agave cantala	Fibers used for making bags.	12
Cannabis sativa	Fibers used for making strings.	96
Gerardenia heterophylla	Fibers used for making strings.	24
Grevia optiva	Fibers used for making strings.	96
Urtica bellebruana	Fibers used for making strings.	12
Bamboos		
Chimnobambusa falcata	Used for making large baskets.	96
C. jaunsarensis	Used for making large baskets, brooms and used as thatch in forest steads and for lighting.	88
Thamnocalamus falconeri	Used in musical instruments: flutes and cigar pipes.	20
T. spathiflorus	Used for making mats and fine baskets.	96
Miscellaneous		
Mixture of *Parmelia nepalensis; P. nilgherrensis; Ramalina subcomplanata and Usnea lucea*	Used in perfumes, flavors, and cosmetics.	—
Pine resine	Livestock rearing.	—
Several fodder species (preferably oaks)	Used as bedding material for livestock and as fertilizer for crops.	100
Several broad leaved species	Used as bedding material for livestock and as fertilizer for crops.	100

* *(Based on Rao and Saxena, 1996)*
? *not known to be used in the last 40 years*
— *not used in the study area*

Sociocultural Traditions and Biodiversity Use

Sociocultural traditions often force a concern of sustainable use of biodiversity. In the northeast, monocropping is restricted to valley lands while mixed cropping is mandatory for shifting agriculture in the uplands. Decisions on allotment of land to households for shifting agriculture, time of slashing, and burning and sowing are taken by the community and not by the individuals. Collective decisions and responsibilities reduce the risks of fires spreading (Ramakrishnan, 1992). In the settled terraced agroecosystems of northwestern and central Himalaya, the agricultural land of a village is divided into two halves. Each half is fallowed once in a two year period (Maikhuri et al., 1996). This tradition favors collective efforts of protecting the crops from damage by wildlife and livestock.

Table 12. Monetary cost and benefit (Rs) at different levels in the marketing of medicinal plants

Parameter	Villagers	Local agents	Wholesale dealers
Cost	118402.90	226013.33	398770.11
Benefit	221987.00	391950.00	514730.00
Benefit/cost ratio	1.87	1.73	1.29

(Source: Rao and Saxena, 1996)

Collection of forest products is organized in groups to minimize the chances of unequal sharing over exploitation and for security from wildlife. Permission for utilization of all types of NTFPs, except for medicinal plants, is granted by the resource rich villages to the resource poor villages. Such village to village concessions are guided by similarity in societal composition and nearness. Within a village, socioeconomic equity matters more than market value of a given product. Market valuation assumes importance when the resources available from a village are supplied to far-off villages or urban centers. Customs favor marketing opportunities of NTFPs to the small holders. A case study in the central Himalaya revealed that, while all sections of the society can make bamboo and plant fibre based handicrafts for their own use, only the poor section of the society (those owning <1 ha) were privileged to sell the products (Rao and Saxena, 1996).

Biodiversity and Economic Benefits

One of the major approaches in developing new drugs from plants is to examine the uses claimed for a traditional preparation. Though many have argued that there is a close relationship between a traditional preparation and a drug obtained from the same plant, data supporting such claims are limited. Farnsworth et al. (1985) found that 74% of plant derived drugs were discovered as a result of chemical studies to isolate the chemical substances responsible for the use of the original plants in traditional medicine. The remaining 26% plant derived drugs did not reveal any correlation between their uses as drugs and the traditional uses of plants from which they were derived. Such a conclusion should be accepted with skepticism because of limited published literature on the traditional knowledge.

An important question to be addressed in the context of the use of biodiversity for the global community is: What share of profits should be ensured to the local communities for conserving biodiversity and for their traditional knowledge that provides clues for modern scientific and technological innovations? Conservation of a pharmaceutically important species would not be at all important when the naturally occurring active principle is chemically synthesized. Yet, the role of traditional knowledge in characterization of the active ingredient cannot be ignored.

At the present, local communities are involved in the production of cash crops and the collection of useful plant products from the wild. The poor economic status of the local communities forces them to sell the products at marginal profits. A significant share of profits from bioresource is realized by the middlemen in the marketing process (Table 12). With increasing impacts of market forces and lack of policy initiatives for strengthening the marketing and value-addition capacity of local people often lead to exploitation regime exceeding the natural regeneration capacity (Farooquee and Saxena, 1996).

Conclusions

Many ecological studies are available on a variation in biodiversity in space and time at varied scales. The spatial-temporal patterns of variation are so complex that prediction of biodiversity value and dynamics with the existing state of knowledge is a difficult task. The hypothesis proposed for biodiversity-ecosystem function relationships needs rigorous testing. Biodiversity is concentrated in the areas dominated by traditional societies. Many traditional uses of biodiversity may provide material and clues for solutions to problems of a wider community. The goal of conservation and sustainable utilization of biodiversity would be achieved when the interests of economically poor traditional societies are paid due attention.✧

Selected Reading

Arroyo, M.T.K., Squeo, F., Armesto, J. and Villagran, C., 1988. Effects of aridity on plant diversity in northern Chile Andes. Annals of Missouri Botanical Garden 75:55–78.

Arroyo, M.T.K., Cavieves, L., Marticorena, C. and Munoz-Sehick, M., 1995. Convergence in the Mediterranean floras in central Chile and California: insights from comparative Biogeography, p. 43–48. In: M.T.K. Arroyo, M.D. Fox, and P. Zedler (eds.), Ecology and biogeography of Mediterranean eecosystems in Chile, California and Austria. Springer-Verlag, New York.

Belsky, A.J., 1990. Tree/grass ratios in East African Savanna: a comparison of existing models. Journal of Biogeography 17:483–489.

Chandrashekara, U.M. and Ramakrishnan, P.S., 1994. Vegetation and gap dynamics of a tropical wet evergreen forest in the Western Ghats of Kerala, India. Journal of Tropical Ecology 10:337–354.

Clark, D.A. and Clark, D.B., 1984. Spacing dynamics of a tropical rain forest tree: evaluation of the Janzen-Connell model. American Naturalist 104:547–559.

Condit, R., Hubbell, S.P., Lafrankie, J.V., Sukumar, R., Manokaran, N., Foster, R.B. and Ashton, P.S., 1996. Species-area and species-individual relationships for tropical trees: a comparison of three 50-ha plots. Journal of Ecology 84:549–562.

Cornell, H.V., 1993. Unsaturated patterns in species assemblages: the role of regional processes in setting local species richness, p. 243–252. In: Species diversity in ecological communities: historical and geographical perspectives. University of Chicago Press, Chicago.

Cowling, R.M. and Holmes, P.M., 1992. Endemism and speciation in a lowland flora from the Cape Floristic Region. Biological Journal of the Linnean Society 47:367–383.

Cowling, R.M., Holmes, P.M. and Rebelo, A.G., 1992. Plant diversity and endemism. In: R.M. Cowling, (ed.), The ecology of fynbos, nutrients, fire and diversity. 11 Oxford University Press, Cape Town 67–112.

Crawley, M.J., 1996. The structure of plant communities, p. 1–50. In: M.J. Crawley (ed.), Plant ecology. Blackwell Scientific Publication, Oxford.

Currie, D.J., 1991. Energy and large-scale pattern of animal and plant species richness. American Naturalist 137:27–49.

Cushman, J.H., 1995. Ecosystem level consequences of species additions and deletions on Islands, p. 135–147. In: P.M. Vitousck (ed.), Islands. Springer Verlag, Berlin.

De Angelis, D.L. and Waterhouse, J.C., 1987. Equilibrium and non-equilibrium concepts in ecological models. Ecological Monographs 57:1–21.

Dobariyal, R.M., Singh, G.S., Rao, K.S. and Saxena, K.G., 1996. Medicinal plant resources in Chhakinal watershed: traditional knowledge, conservation and economy. Journal of Herbs, Spices and Medicinal Plant, 4. (in press).

Ehrlich, P.R. and Ehrlick, A.H., 1981. Extinction. The causes and consequences of the disappearance of species. Random House, New York.

Farnsworth, N.R., Akerele, O. and Bingel, A.S., 1985. Medicinal plants in therapy. Bulletin of the World Health Organization 63:965–981.

Farooquee, N.A and Saxena, K.G., 1996. Conservation and utilization of medicinal plant in high hills of central Himalaya. Environmental Conservation 23:75–80.

Franklin, J.F. 1993. Preserving biodiversity: species, ecosystems or landscapes. Ecological Applications 3:202–205.

Gentry, A.H., 1988. Changes in plant community diversity and floristic composition on environmental and geographical gradients. Annals of the Missouri Botanical Garden 75:1–34.

Grime, J.P., 1973. Control of species density in herbaceous vegetation. Journal of Environmental Management 1:151–167.

Groves, R.H. 1994 (Ed). Australian Vegetation. 2nd Edition, Cambridge University Press., Cambridge.

Grubb, P.J., 1977. The maintenance of species-richness in plant communities: the importance of the regeneration niche. Biological Reviews 52:107–114.

Heywood, V.H. and Watson, R.T., 1995. Global Biodiversity Assessment. Cambridge University Press, Cambridge.

Hubbell, S.P. and Foster, R.B., 1986. Biology, chance and history and the structure of tropical rainforest tree communities, p. 314–319. In: J. Diamond and T.J. Case (eds.), Community Ecology. Harper and Row, New York.

Hulme, P.E., 1996. Herbivory, plant regeneration, and species coexistence. Journal of Ecology 84:609–616.

Janzen, D.H., 1970. Herbivores and the number of tree species in tropical forests. American Naturalist 114:501–528.

Jeltsch, F., Milton, S.J., Dean, W.R.J. and Rooyen, N.V., 1996. Tree spacing and coexistence in semiarid savannas. Journal of Ecology 84:583–595.

Kohn, D.D. and Walsh, D.M., 1994. Plant species richness—the effect of island size and habitat diversity. Journal of Ecology 82:367–377.

Latham, R.E. and Ricklefs, R.E., 1993. Global patterns of tree species richness in moist forests: energy-diversity theory does not account for variation in species richness. Oikos 67:325–333.

Levin, S.A., 1974. Dispersion and population interactions. American Naturalist 108:207–228.

Lieberman, D., Lieberman, M., Peralta, R. and Hartshorn, G.S., 1996. Tropical forest structure and composition on a large scale attitudinal gradient. Journal of Ecology 84:137–152

Lovejoy, T.E., 1995. The quantification of biodiversity: an esoteric quest for or a vital component of sustainable development, p. 81-88. In: D.L. Hawksworth (ed.). Biodiversity Measurement and Estimation. The Royal Society and Chapman and Hall, London.

MacArthur, R.H. and Wilson, E.O., 1967. Theory of Island Biogeography. Princeton University Press, Princeton.

Maikhuri, R.K., Rao, K.S. and Saxena, K.G., 1996. Traditional crop diversity for sustainable development of Central Himalayan agroecosystem. International Journal of Sustainable Development and World Ecology 3:8–31.

Menge, B.A., Berlow, E.L., Blanchette, C.A., Navarrete, S.A. and Yamada, S.B., 1994. The keystone species concept: variation in interaction strength in a rocky intertidal habitat. Ecological Monographs 64:249–286.

Myers, N., 1988. Threatened biotas: "hot spots" in tropical forests. The Environmentalist 8:187–208.

Myers, N., 1990. The biodiversity challenge: expanded hot-spots analysis. The Environmentalist 10:243–256.

Pacala, S.W., 1986. Neighbourhood models of plant population dynamics. 4. Single-species and multispecies models of annuals with dormant seeds. American Naturalist 128:859–878.

Pacala, S.W. and Roughgarden, J., 1982. Spatial heterogeneity and interspecific competition. Theoretical Population Biology 31:92–113.

Paine, R.T., 1969. A note on trophic complexity and community stability. American Naturalist 103:91–93.

Power, M.E. and Mills, L.S., 1995. The keystone cops meet in Hilo. Trends in Ecology and Evolution 190:182–194.

Prance, T.G., 1995. A comparison of the efficacy of higher taxa and species numbers in the assessment of biodiversity in the neotropics, p. 89–99. In: D.L. Hawksworth (ed.), Biodiversity measurement and estimation. The Royal Society and Chapman and Hall, London.

Ramakrishnan, P.S., 1992. Shifting Agriculture and Sustainable Development: an interdisciplinary study from north-eastern India. Parthenon Publishing Group, Park Ridge.

Ramakrishnan, P.S., Purohit, A.N., Saxena, K.G., Rao, K.S. and Maikhuri, R.K. (eds.), 1996. Conservation and management of biological resources in Himalaya. Oxford and IBH, Delhi.

Rao, K.S. and Saxena, K.G., 1996. Minor forest products management— problems and prospects in remote high altitude villages of central Himalaya. International Journal of Sustainable Development and World Ecology 3:60–70.

Rao, R.R., 1996. Hot spots of biodiversity in India. In: P.S. Ramakrishnan, A.K. Das and K.G. Saxena (eds.), Managing biodiversity for sustainable development. Indian National Science Academy, New Delhi (in press).

Ricklefs, R.E., 1977. Environmental heterogeneity and plant species diversity: a hypothesis. American Naturalist 111:376–381.

Ryvarden, L., 1993. Tropical Polypores, p. 149-180. In: S. Isaac, J.C. Frankland, R. Watling, and A.J.S. Whalley, As-pects of tropical mycology. Cambridge University Press, Cambridge.

Sale, P.F., 1977. Maintenance of high diversity in coral reef fish communities. American Naturalist 111:337–359.

Saxena, K.G. and Ramakrishnan, P.S., 1983. Growth and allocation strategies of some presnnial weeds of slash and burn agriculture (Jhum) in north-eastern India. Canadian Journal of Botany 61:1300–1306.

Saxena, K.G. and Ramakrishnan, P.S., 1984. C_3/C_4 species distribution among successional herbs following slash and burn in northeastern India, p. 335–346. Acta Oecologia/Oecologia Plantarum.

Shugart, H.H., 1984. A Theory of forest dynamics. Springer-Verlag, New York.

Singh, G.S., Rao, K.S. and Saxena, K.G., 1996. Energy and economic efficiency of the traditional mountain farming system: a case study in north-western Himalayn. Journal of Sustainable Agriculture (in press).

Swaine, M.D. and Whitmore, T.C., 1988. On the definition of ecological species groups in tropical rain forests. Vegetation 75:81–86.

Tilman, D., 1982. Resource competition and community structure. Princeton University Press, Princeton.

Tilman, D., 1994. Competition and biodiversity in spatially structured habitatas. Ecology 75:2–16.

Tilman, D. and Pacala, S.W., 1993. The maintenance of species richness in plant communities. In: R.E. Ricklefs and D. Schluter (eds.), Species diversity in ecological communities. University of Chicago Press, Chicago.

Toumisto, H., Ruokolainen, K., Kalliola, R., Linna, A., Danjoy, W. and Rodriguez, Z., 1995. Dissecting Amazonian biodiversity. Science 269:63–66.

Turkington, R. and Joliffe, P.A., 1996. Influence in *Trifolium repens-Lolium perenne* mixtures: short- and long-term relationships. Journal of Ecology 84:563–572.

Valencia, R., Balslev, H. and G.Pazymio, 1994. High tree alpha diversity in Amazonian Ecuador. Biodiversity and Conservation 3:21–28.

Vitousek, P.M., Walker, L.R., Whiteacre, L.D., Mueller-Dombios, D. and Matson, P.A., 1987. Biological invasion by *Myrica faya* alters ecosystem development in Hawaic. Science 238:803–804.

Walker, B.H., 1985. Structure and function of savannas: an overview, p. 83-91. In: J.C. Tothill and J.J. Mott (eds.), Ecology and management of the world's savannas. Australian Academy of Science, Canberra.

Walker, B.H., 1992. Biodiversity and ecological redundancy. Conservation Biology 6:18–23.

Warner, R.R. and Chesson, P.L., 1985. Coexistence mediated by recruitment fluctuations: a field to the storage effect. American Naturalist 125:769–787.

Whittaker, R.H., 1972. Evolution and measurement of species diversity. Taxon 21:213–251.

Wiens, J.A., 1984. On understanding a non-equilibrium world: myth and reality in community patterns and processes. In: D.R.Strong, D. Simberloff, L.G. Abele and A.B. Thistle (eds.), - Ecological communities: conceptual issues and the evidence. Princeton University Press, Princeton.

Biodiversity and Biotechnology: Ethical Issues in Developing Countries

Suresh K. Sinha and Suman Sahai
Water Technology Centre, IARI, New Delhi, India;
Gene Campaign, New Delhi, India

DEVELOPMENTS IN BIOTECHNOLOGY have brought a phase in the intellectual property regime (IPR) which biological scientists and the scientific community in the developing countries were not prepared for. Biodiversity constituted the foundation of crop improvement in national and international programs which benefitted all nations including the developing and developed countries. It seems a dream that the international centers such as the International Center for Corn and Wheat (CIMMYT) and the International Rice Research Institution (IRRI) could collect germplasm from everywhere and whatever was required or wanted was available without any reservation or Memorandum of Understanding (MOU). To the credit of these institutions, they developed genotypes of wheat and rice which were utilized directly or through further modification to usher in the "Green Revolution." It is conceivable that many of the developing countries in the absence of a green revolution would have had a red revolution. However, the developed countries of Europe and North America and others also benefitted by exporting food grains and other agricultural products. Thus the biodiversity of crops became truly a common heritage of mankind and helped everyone.

The basic philosophy of the East and West has been quite different. Even when monarchism flourished in the East, the monarchs (such as Ashok and Akbar) who believed and practiced "Sub Jan Sukhay" or "Welfare to All" have been revered, and those who renounced their kingdom, like Buddha and Gandhi, are worshipped. Underlying this philosophy is the concept of social welfare rather than only welfare for the self. For some reasons the West developed the philosophy self and subjugation of others by any means. Some people attribute it to Darwin's theory of evolution underlining the concepts of struggle for existence and survival of the fittest. Thus, improvement of the self, including personal benefit rather than public good, became important objectives of Western society. There is nothing wrong in it if it is not done by using the resources of others. Since moral and ethical objectives imply the safeguards of the public at large, we have started having discussions on these issues now.

The most important human need of food has been met through agriculture and will continue in the future. With the increasing population, the necessity of improvement of crops (other commodities also) was essential and took the form of a production system in the West because it followed the pattern of the industrial revolution and industrial production system—few people, large investment, high production, and export and profit generation. Thus "seed" as an input in the production system had to have a value, and accordingly "seed" was to be an industry. Improvement of "seed" required diversity, collection, and utilization. The farmers in the East had no hesitation in exchange of the seed. There used to be villages which were known to grow some crops more efficiently than others, and the seed developed in one village was available to others. Wherever agricultural research institutions grew up, their mandate was to provide "seed" to farmers through public institutions. The latter has not necessarily been an efficient system but has served an important cause of public service. The green revolution in India was the product of many inputs and policy actions, but "seed" was provided by the public seed agencies. All this goes to show that in India and other developing countries, the concept of patenting of biological material did not arise. Therefore, the rice collections in thousands from Central-East India and North-East India and other parts were provided to the IRRI. Some of these cultures provided resistance to insects, virus, and other diseases. Today these collections are available to anyone–the public or private sector. The genes for these traits could be identified and isolated and patented. The same way the germplasm of pigeonpea, pearl millet, Sorghum, Chickpeas, and groundnut was transferred to the International Centre of Research on Semi-Arid Tropics (ICRISAT), Hyderabad. We believe that this was right and has helped humanity.

A change from this scene is now taking place. Earlier, working in a famous laboratory with a famous scientist entailed no restrictions. Everyone worked together. Now there are instances of signing a MOU when a famous laboratory wanted a visiting scientist sponsored by our institute through the United Nations Development Program (UNDP)-Food and Agricultural Organization (FAO). This stated that the visiting scientist will leave all original records and take no photocopies and information. This was to be signed by the Director of the sponsoring institute, providing all financial help through its own programs. We also have seen some institutions not explaining what they were doing. Such institutions do not allow visitors to visit certain laboratories. Science has become secretive. This prompts a similar response from others also. For example, one of us saw a lady threshing *Brassica campestris* in a remote area. We got the seeds from her and tested for dwarfness. The seeds proved true to the collected material. Normally such a material would be distributed to a large number of persons to make a quick advance towards crop improvement. But there is a dilemma. The lady who had the plants did not know its value to others. The istitute collected it and confirmed its potential. If given to others and the private sector, would they benefit this small hamlet from where the seed was collected? What kind of benefit? The seed has not been distributed and the research continues at a slow pace. This is a loss to science and to the global crop improvement program.

In another case, an educated and progressive farmer saw plants with large spikes in the wheat. He selected them and continued to further select, leading to development of a new wheat variety. Similar observations were made in cotton out of the seed obtained from seed agencies or research institutes. The farmer makes seeds every year and markets them to farmers of the region. He gets a nice profit from the sale of his seeds. It is possible that one may find variants in seeds obtained from either public or private sector organizations. Does it involve any ethical or moral issues? Perceptions would differ among the public sector, private sector, scientists, individuals, and farmers, but these will be the problems in developing countries.

Some of the examples shown above demonstrate the importance of biodiversity. The necessities and demands of man have created a greater need for biodiversity for the public good. No new gene or promoter has been synthesized that does not exist in nature though theifunctioning and methods of use have been developed. Biotechnology has opened an avenue for the betterment of humanity, but secrecy and commercialization are leading to suspicion and the potential for monopolies. Scientists, like others, also look for comforts. There may be unethical practices, malpractices, and legal battles. This may lead to unethical practices and impoverishment of science. It is, therefore, necessary to maintain openness.

A new debate in India is on the evils of biotechnology. Running parallel to this is the debate on bioethics. Both are marked by their plagiarized western rhetoric. The metaphors are lifted straight from the radical fringes of Europe and repeated almost verbatim on Indian platforms.

The violent objections to biotechnology would appear to have a logical basis for the social context and economic situation in western societies. In the case of Germany, the resistance to genetics must be seen in the context of the Third Reich and the reprehensible eugenics program of the Nazis. The fact is that all these have societies that are not only self sufficient in food, but also have a standard of food availability and choice that perhaps cannot be bettered.

Europe and America not only have a very high level of food security, they produce such volumes of surplus that it costs money to destroy the mountains of fruit and vegetables, the lakes of milk and wine, and the stacks of meat and butter. In 1993, it cost the Europeans over 3.6 mio DM to destroy the surplus fruit and vegetables that could not be consumed or processed. Why would these societies welcome a biotechnology route to produce still more food?

Western concerns focus on genetics violating the dignity of man. We hear of the dehumanizing impact of gene technology and the terrifying dangers of biotechnology in agriculture, dairy, and food production. But can we in India have the same perception? Is it more unethical to "interfere" in God's work or is it more unethical to allow hunger deaths when these can be prevented?

In India where post-harvest losses destroy up to 30% of the food we bring in from the field, should western hysteria about biotechnology be allowed to get in the way of making agricultural products more durable and amenable to processing? Should 50% and more of the fruit grown in the economically weak hill regions be allowed to rot because it cannot be sold or should we try to produce fruit varieties which can be processed to delay rotting?

The issue of biotechnology and bioethics is often confused with that of biosafety in the sometimes incoherent debate on genetics and genetics application. The crucial importance of biosafety in genetics and biotechnology cannot be overemphasized.✧

Ethics and Biodiversity

Niraja Gopal Jayal
Center for Political Studies, Jawaharlal Nehru University, New Delhi, India

Introduction

The chief policy and indeed political questions that have been at the center of debates on biodiversity and biotechnology in recent times are underpinned by assumptions and convictions of a strongly ethical nature. It would not, in fact, be an exaggeration to assert that at the bottom of every policy and political statement on the subject there lies implicit a normative position about the relationships between nature and human beings, as well as the diverse social, political, and moral communities to which individuals belong. The first part of this paper frames the discussion of ethics and biodiversity by positing a distinction between two broad categories of arguments relating to the environment and relating these to the major policy positions that enjoy currency today. In the broadest terms, these two blanket categories (that conceal a variety of philosophical differences) may be characterized as arguments of an ecocentric nature *versus* those of an anthropocentric nature. The second part of the paper identifies and problematizes three specific issues in the debate on biodiversity–those of national sovereignty, local communities, and future generations.

The essential characteristic of all ecocentric arguments—as opposed to anthropocentric ones—is the claim of intrinsic value, which encourages us to respect nature for it's own sake, rather than for human purposes. The claim of intrinsic value is, therefore, opposed to an instrumentalist view of nature. Ecocentric arguments derive intellectual sustenance from a variety of philosophical traditions but preeminently from Utilitarian and Kantian ethics. Furthermore, in the contemporary discourse on the environment, they have some affinity with deep ecology. These arguments, however, have a limited potential for generating concrete policy strategies in relation to the political, economic, and cultural questions that are indisputably at the center of the debate on biodiversity and biotechnology today.

Hedonist utilitarian ethics, for instance, makes *sentience* the criterion for identifying the defensible boundaries of concern for the interests of others. This was precisely the criterion used by Jeremy Bentham in the nineteenth century with respect to the question of slavery in the British dominions. "What matters," said Bentham, "is not whether they can reason or talk, but whether they can suffer." A capacity for suffering—and its obverse, a capacity for enjoyment and happiness—gives a being a right to equal consideration because it suggests that the being in question has *interests*. A stone cannot suffer. Hence, it does not have interests, and we have no moral obligation to concern ourselves about its welfare. But there is no moral justification for not taking into consideration the interests of beings who do suffer. Regardless of the nature of the being, the principle of equality requires that its suffering be counted equally with the like suffering of any being (Singer, 1986). Of course, it is arguably not necessary to be a utilitarian in order to recognize the moral significance of pain. (Johnson, 1993:50)

While these ethical considerations can be extended beyond humans to sentient beings, such as animals, can they similarly be extended to other living things? Plants and trees cannot, after all, be said to have wants and desires of their own, or indeed the capacity to suffer. In classical Greek and Roman philosophy (especially in the writings of Aristotle and Thomas Aquinas) we frequently encounter the metaphor of the tree as it grows and develops according to its own inner nature. Does the flourishing or languishing of plants suggest a good in itself—an intrinsic value? If we do extend our respect for sentient beings to nonsentient living things, on what basis do we differentiate between the greater or lesser worth of some in situations where practical tradeoffs have to be made?

An influential response to these questions has been that of deep ecology, first enunciated in Arne Naess's well-known distinction between "shallow" and "deep" strands in the ecological movement. As opposed to shallow ecology which is limited to traditional moral frameworks and encourages conservation for the enjoyment of human beings, deep ecologists wish to preserve the integrity of the biosphere *for its own sake*, regardless of any benefits to human beings that might flow from this.

The shallow ecology movement thus fights against pollution and resource depletion, inspired by the central objective of concern for the health and affluence of people in the developed countries (Naess, 1991). Deep ecology, by contrast, supplants the man-in-environment image by a total-field model. This posits an intrinsic relation between things, such that the relation is itself constitutive of the definition of the things in question. It also accords respect to ways and forms of life, on the premise that the equal right to live and blossom is a value axiom. This is the basis of the principle of biospherical egalitarianism, which forces us to recognize our close partnership and interdependence with other forms of life. It leads, further, to a recognition of the principles of diversity and richness of forms and modes of life, replacing the principle of the survival of the fittest with that of coexistence and cooperation of different ways of life, cultures, and economies. The value accorded to diversity translates in social terms into an anticlass position and a politics of decentralization and local autonomy. (ibid.)

For deep ecologists, thus, species, ecological systems, and the biosphere, as a whole, are objects of value. They value human and nonhuman life on earth as having intrinsic value; they believe that the richness and diversity of lifeforms contributes to the realization of this value; and they are convinced, finally, that human beings have no right to diminish this richness and diversity except for the satisfaction of vital needs. This argument has been extended to assert that all organisms and entities in the ecosphere, being parts of an interrelated whole, are equal in their intrinsic worth. One problem with this argument is, of course, that in privileging the worth of the whole ecosystem, it provides no grounds from which the value of individual plants or microorganisms might be determined. All organisms may be part of an interrelated whole, but this is insufficient to establish that they are all of intrinsic worth, let alone of equal intrinsic worth. (Singer, 1986:282)

Ecocentric arguments rooted in an essentially Kantian perspective justify the moral significance of the interests of the biosphere in a way that addresses this gap. Kantian ethics, as is well known, calls upon us to treat others as ends in themselves, and never only as a means to our ends. It has been argued that this implies that what is good for a person must be determined by that person's own nature and self-identity. Extending Kantianism in this fashion makes it capable of accommodating not only rational beings as objects of moral concern but others as well. Kantianism, in fact, provides grounds for establishing and justifying the moral significance of the interests of species and ecosystems. Like individuals, species have an interest in surviving well in fulfilling their nature in an appropriate environment. A species is a living system, an ongoing coherent organic whole, with properties that are not simply the aggregate of those of the individual species members. Similarly,

> "The biosphere is not the sum of all living beings, nor yet is it a separate entity. It has interests that are not entirely separate, yet which are not the aggregate interests of the various living entities. And contrary to what some extreme holists might suggest, not just the interests of the biosphere are morally significant. Yet, if we follow out the implications of taking species as morally significant entities, we certainly do come to the conclusion that the biosphere is a morally significant entity" (Johnson, 1993:265).

If ecocentric arguments enjoin an essentially negative, leave-alone policy attitude towards biodiversity, anthropocentric arguments mandate what might be characterized as a positive strategy towards its conservation. The best philosophical bridge between ecocentric and anthropocentric arguments is provided by an ethic that appeals to the Aristotelian conception of well-being—

> "...According to which well-being should be characterized, not in terms of having the right subjective states as the hedonist claims, nor in terms of the satisfaction of preferences as modern welfare economics assumes, but rather in terms of a set of objective goods a person might possess (i.e., friends) the contemplation of what is beautiful and wonderful, the development of one's capacities, the ability to shape one's own life, and so on" (O'Neill, 1993:3).

This approach claims to go beyond the narrowly anthropocentric to argue that we value things in the natural world for their own sake and not simply as an external means to our own satisfaction. The relationship between the natural world and human well-being is seen as profoundly similar to that between friendship and human flourishing in Aristotle.

> "It is constitutive of friendship of the best kind that we care for friends for their own sake and not merely for the pleasures or profits they might bring. To do good for friends purely because one thought they might later return the compliment, not for their own sake, is to have an ill-formed

> friendship. Friendship, in turn, is a constitutive component of a flourishing life. Given the kind of beings we are, to lack friends is to lack part of what makes for a flourishing human existence" (ibid:24).

This argument entails, in terms of policy prescription, an Aristotelian rejection of the market—because this encourages a view of well-being that is highly parasitical upon the unceasing acquisition of consumer goods—in favor of nonmarket associations which provide the most favorable context for fostering the well-being of human persons, the nonhuman world, and also future generations.

The anthropocentric argument, in its best and broadest sense, makes possible the consideration of three issues which have assumed much political importance in recent debates on biodiversity and biotechnology. These issues are:

- The question of national sovereignty
- The question of local communities
- The question of future generations

Each of these issues is underwritten by essentially liberal principles of autonomy and agency, with the possible exception of the last which can logically contain only an implicit notion of potential agency. In each case, though in very different ways, the claim of autonomy generates a claim to rights. Thus, the claim of national sovereignty translates very smoothly into the right of the sovereign nation to control and manage natural resources within its territorial boundaries as it wishes. Likewise, the claim of autonomy for local communities generates the idea of their rights—as collectivities—to the ecosystem which not only sustains them, but to the preservation and improvement of which they have historically contributed through creative practices now recognized as indigenous knowledge systems. Finally, the potential for agency and autonomy that is attributed to future generations (on the assumption of their continuity with the present) mandates the recognition of their rights to a natural world that is not destroyed and eroded either in the quantum of its resources or in their richness and diversity. Each of these claims, as we know, further translates from the ethical to the political realm, to yield definite policy prescriptions.

It is imperative also to recognize that each of these claims takes the particular form it does—that of a *claim*—precisely because there are serious economic and political constraints on their fulfillment. It is not merely the awareness of depleting resources that has impelled this reaction, but to a substantial degree, the predatory behavior of powerful states of the North as well as transnational corporations. Let us examine each of these issues individually, in relation both to the context in which they have found articulation, and also in terms of their moral and philosophical status.

The Question of National Sovereignty

The chief philosophical issue underlying this question is that of who should own and control natural resources (in this case, resources rich in biodiversity) and, by extension, who should determine how these resources may be used. Do these resources belong to humankind in general, to nation-states within whose territory they happen to subsist, or to the peoples who live in and by them?

On the answer to this first question depends the answer to the further questions of both the limits of use and also of the infringement of national sovereignty. It is, for instance, possible to argue that such resources belong to humankind in general, and that there is, therefore, nothing wrong in countries exploiting for commercial purposes strains which have originated in other countries. This argument has the spurious attraction of being consistent with the rhetoric of global unity and the common endeavors of mankind, but its moral appeal is quickly undermined when these uses are sought to be patented in the form of intellectual property rights in the context of a trade regime that is patently unequal as between nations. Thus, do the developed nations which have the material and intellectual resources to experiment and devise adaptations in sophisticated laboratory conditions, also have the moral right to all future use of those genetic resources, regardless of the country of their origin that is frequently also a developing nation? This question is especially charged because many countries of the South, including India, are rich in germplasm while Northern countries which have made great advances in technology in the field of genetics and genetic engineering have little or no germplasm (Sahai, 1995). It may be useful to consider these issues against the backdrop of the international conventions on this question.

As is well known, the question of biodiversity conservation received its earliest expression in the form of the concern for the destruction of the tropical rainforests, articulated mostly by Northern nongovernment organizations (NGO)s like the World Wildlife Federation (WWF) and the International Union of

Conservation and Nature (IUCN). In collaboration with the World Resources Institute, the World Bank, and the United Nations Environmental Programme, it was these NGOs which drafted the original documents on biodiversity conservation. These drafts reflected a fairly traditional environmentalist approach–that of resource management. It was only in 1991 at a Geneva meeting that the Group of 77 countries asked for the issue of biotechnology to be included with that of biodiversity. A month before the UN Conference on Environment and Development (the Earth Summit) held at Rio de Janeiro in June 1992, a document was drawn up at Nairobi that embodied substantial concessions for which the United States (U.S.) had driven a hard bargain. However, the U.S. eventually failed to sign the Biodiversity Convention at Rio. The Convention on Biological Diversity, in force since December 1993, however, has been ratified by 127 countries.

The general declaration on the environment and development that emerged from the Rio Summit reflects an unambiguously anthropocentric approach towards sustainable development. It also gives to nation-states the "sovereign right" to "exploit" their natural resources, in accordance with their own policies on environment and development. Given the further recognition of a right to development—subject only to the minor caveat of the developmental and environmental needs of future generations—it communicates the priority of development over the environment, rather than the idea that development should take place only within environmental limits.

While it is true that many Southern countries assert national sovereignty only to pursue an imitative path of development, even environmental activists are frequently forced to resort to nationalist arguments in the face of threats from a predatory global political economy. Chatterjee and Finger have argued that the biodiversity convention exemplifies a perversion of the concern for the destruction of the world's biodiversity into a preoccupation with new scientific and biotechnological developments for economic growth. There are, in their view, the following three key arguments which hold this perversion together:

> First, the convention gives "nation-states the sovereign right to exploit their own resources pursuant to their environmental policies," thus transforming biological diversity into a natural resource to be exploited and manipulated. Then, the convention implicitly equates the diversity of life—animals and plants—to the diversity of genetic codes, for which there

> are genetic resources. By doing so, diversity becomes something modern science can manipulate. Finally, the convention promotes biotechnology as being "essential for the conservation and sustainable use of biodiversity" (Chatterjee and Finger, 1995).

Consequently, biotechnology comes to be projected as something which is unambiguously good for the conservation of biodiversity, as it is conducive to the maintenance of genetic diversity, as well as for improving agricultural production, through making possible the increasing productivity of crops, livestock, and aquaculture species. Biotechnology, on this account, becomes the ideal instrument through which to secure not only material progress, but also biodiversity conservation. That this is not an altogether innocent agenda, of biotechnologists laboring painstakingly for the common future of mankind, is well-known. That there are also significant commercial interests—most notably in the pharmaceutical and biotechnology industries—to be served by formalizing such arguments in international conventions is equally well-known.

As biotechnology has moved out of the cloistered world of scientific laboratories into the market, the stakes have changed—

> Biotechnology is slated to account for almost 60 to 70 percent of the global economy for at least the next two to three decades. What is of special relevance is that biotechnology covers a span of economic sectors which is unprecedented. It will play a role in fields as diverse as mining, feedstock chemicals, energy, pharmaceuticals and, of course, food. Disconcertingly, going by present trends, this biotechnology sector will be almost entirely in the hands of the 10 to 12 largest multinationals of the world (Sahai, 1995).

There still remains something to be said for the argument that this controversy has gotten embroiled in secondary conflicts of a financial and political nature, quite forgetting the primary question of what is causing the destruction of biodiversity and what can be done about this. In allowing sovereign governments (and, by implication, local industries) the right to use and manipulate their own natural resources, it also loses sight of another important issue (the communities who depend on biodiversity for their sustenance, their habitat, food, medicines, and even culture).

The Question of Local Communities

In ethical terms, this question has two components. The first is the argument that the local communities who live in and by these natural resources have an historically validated right to them, and the second is that these communities are living repositories of knowledge systems about these biological resources, on the basis of which they have creatively adapted and innovated in impressive ways that deserve recognition.

The recognition of the rights of local communities—including but not always indigenous peoples—became an issue in recent years, chiefly in the context of involuntary displacement caused by development projects like big dams or experiments in forest management. The claim of such rights signalled a radical departure from the conventional liberal premise that rights-bearing agents are typically individuals. Indeed, the admissibility of claims to collective rights became linked with similar claims made by cultural communities such as indigenous peoples in Canada and Australia. Can communities have rights, and if so, on what basis? In recent years political theorists have come to agree that they can and that such rights are admissible even within liberal political philosophy, not just communitarianism that is naturally hospitable to such claims.

Establishing the admissibility of rights-claims for collectivities, and not just for individual citizens, entails recognizing that communities, no less than individuals, possess a capacity for agency and autonomy. Though classical liberal theory attributed this capacity exclusively to individuals—often defined in such a way as to exclude even colonized peoples—the contemporary discourse on rights gives explicit recognition to such claims. If it is, then, in principle possible for collectivities and communities to have rights, what is the substance of such rights in relation to biodiversity? The basis of the claim is chiefly historical as it invokes the customary lien such communities have enjoyed over these resources. It also appeals to an ideal of pluralism and diversity to suggest that the diversity of human cultures parallels the diversity of the natural world and that the philosophical justification for the preservation of both is not very dissimilar.

Do communities that have lived in and by biodiversity, for whom biodiversity has supplied not only all their material needs (e.g., food, fodder, house construction material, and raw material for medicines) but also their cultural identity, have any rights over those resources? Does the fact of their having been historical inhabitants, users, and conservers of these resources not give them a special lien on these, which overrides the predations of the state and industry? The polar opposite of this argument is, of course, explicitly developmental. This suggests that development, which is equated with the material progress of the nation, is hampered and obstructed by such claims by or on behalf of local communities. Another variation of this argument is that of "mainstreaming" tribal communities, on the premise that they are being excluded from and, therefore, deprived of the benefits of progress and development by being kept caged within their traditional unchanging environs. What these arguments ignore is both the abstract right of local communities to the natural resources which have historically provided them with their sustenance, as well as the contributions of these communities in terms of not only the conservation of these resources but also the innovations they have brought about through their own creativity and their intimate knowledge of the properties of various species.

This is the basis for the further claim to recognition for the practices and creative innovations of these communities, as they have historically interacted with their natural environment. (For a rich variety of examples, see Warren et al., 1995). This implies that, as the authors of these innovations, they possess a sort of copyright over them, that must be respected against egregiously invasive attempts at intellectual piracy. The claim to collective or community rights thus translates effortlessly into a claim to collective intellectual property rights. This claim has a certain normative charge, which similar claims made by transnational corporations and sought to be enshrined in multilateral agreements between nations lack to the extent that they reflect self-interest, whether national or commercial.

The claim of communities to indigenous knowledge and innovations is bereft of the connotations of commoditization and potential commercialization that the international attempts to legislate on this subject clearly are not. It is on this ethical basis that attempts have been made to assert collective intellectual property rights to counter a manifestly unfair and iniquitous patents regime. These have on occasion taken the form of the political mobilization of farmers, for instance, to protest against the Dunkel Draft or the announcement that patenting "neem" is an act of intellectual piracy by companies who seek to use local knowledge and resources without permission.

Thus, while the sovereign rights of nation-states to biodiversity have received international recognition in the form of the Biodiversity Convention, these rights need to be further translated into the *prior* rights of the communities who have maintained and preserved it. It would, however, require an exceptionally enlightened national government to attempt to do this through relevant legislation and to refrain from predatory interventions, whether on its own behalf or by industry and market forces. In the post-Rio era, the onus lies squarely on governments to ensure that the diversity of both species, as well as of communities, is protected. Vandana Shiva has argued persuasively that there are three imperatives here: (a) the ethical and ecological imperative of recognizing the intrinsic worth of all species; (b) the imperative of giving equal recognition to creativity in diverse cultures, especially indigenous traditions of knowledge; and (c) the economic imperative to provide basic standards of health and nutrition to all citizens (Shiva, 1996). The legal mechanisms through which some or all of these objectives could be secured remain elusive, though attempts are apparently being made to achieve this end.

The Question of Future Generations

Finally, there is the question of our duties towards future generations and of their rights against us. Modern liberal philosophy endorses the principle that the beings who can have rights are precisely those who have or can have interests. But it questions the assumption that plants and vegetables have interests which can translate into rights to protection and conservation. As living things with certain inherited biological propensities which determine the pattern of their natural growth, plants can be said to be capable of having a "good." Is not that "good," however, distinct from having interests? Interests presuppose cognitive equipment. They are "compounded out of *desires* and *aims*, both of which presuppose something like *belief*, or cognitive awareness." (Feinberg, 1980) If the plant's need for nutrition or cultivation, its flourishing or languishing, suggest that it has interests, the interests that thrive when plants flourish are not plant interests, but human interests. However, though neither plants nor whole species can be said to have rights in the strict sense of the term, we may assert duties to protect threatened and endangered species, "not duties to the species themselves, but rather duties to future human beings, duties derived from our housekeeping role as temporary inhabitants of this planet" (ibid.:172).

This *normative* claim, of course, is generally justified by reference to the *fact* of the exhaustibility of natural resources, including but not only biodiversity. Despite a variety of methodological problems in calculating the economic value of biodiversity, it has been argued that in the course of the twenty-first century, the projected loss of species may be to the tune of 20–50% of the world's totals, which means a rate between 1,000 to 10,000 times the historical rate of extinction. Further, it is argued, the rate of loss far exceeds the regenerative capacity of evolution to throw up or evolve new species. Thus, extinction "outputs" far exceed the speciation "inputs" (Pearce and Morgan, 1994).

The implications of species depletion for the integrity of many vital ecosystems are far from clear. The possible existence of depletion thresholds, associated system collapse, and huge discontinuities in related social cost functions are potentially the worst outcome in any reasonable human time horizon. Such scenarios are indicative of the links between ecosystem integrity and economic well-being. More immediately, the impoverishment of biological resources in many countries also might be regarded as an antecedent to a decline in community or cultural diversity, indices of which are provided in diet, medicine, language, and social structure (Pearce and Morgan, 1994).

At one level, it is possible to make the claim that biotechnology enhances the prospects of future generations, as a result of the advances it has made in the area of genetic engineering. At another, however, it is equally validly possible to argue the value of preserving the natural world from further depredations and degradation so that future generations may enjoy it. The first argument is justified by reference to the Enlightenment view that science and material progress have the potential of solving most human problems and that this potential only increases as the frontiers of science expand. The second is justified by a view that is clearly ethicophilosophical as well as ecological.

Any claim for the conservation of biodiversity which appeals to our concern for future generations is, of course, clearly located in an anthropocentric ethic. A deep ecological position on the biosphere is manifestly more difficult to defend on ethical grounds than a human-centered ethic which invokes our belief that our descendants (will) have interests that are morally significant.

Even an aesthetic justification for preserving the biosphere is, at bottom, a human-centered one. Whether it is the material interests of future generations that we wish to protect in the face of an alarming depletion of biodiversity or a consideration for their aesthetic fulfillment that impels us to preserve the wilderness for their enjoyment, the rights we seek to protect for them are "contingent" rights; i.e., the interests that they are sure to have when they come into being (Feinberg, 1980).

The discussion so far suggests that the claims to rights of nation-states and local communities are premised on the assumption of autonomy and agency, while the same are attributed to future generations who are expected to be not unlike us in this matter. It is possible, however, to discern a tension *between* some of these claims which sometimes seem to be less than compatible with each other. This tension raises questions which are difficult to resolve unequivocally. Thus, we assert the sovereign right of the nation-state to the biodiversity resources found within its territorial boundaries and seek to formalize this ownership in instruments of international law. Developing countries have been partially successful in their efforts to this end. While they have secured recognition of their sovereign rights to use and exploit resources (however repugnant that usage is to the environmentalist's sensibility), their attempts at negotiating a less iniquitous international patents regime have not been as efficacious. Thus, what they have secured is the guarantee that sovereign states have the right to use and exploit their natural resources and to set their own terms and conditions on the transfer and sale of genetic resources. What is assumed, but left unlegislated, however, is the rights of local communities to the biological resources by which they live, free from the depredations of both the state, transnational, and indigenous corporate interests. Is the sovereignty given to nation-states (by, for instance, the Biodiversity Convention) a sufficient guarantee of the interests of local communities? There are inherent dangers in vesting excessive faith in national governments. India, for instance, has a none too distinguished record of conservation or sustainable development. It has largely followed a developmental path that is imitative and becomes increasingly so as the economic reforms process (set in motion five years ago) gathers momentum. If the overall economic—and, therefore, developmental—strategy is so imitative, what price are the assurances of the Biodiversity Convention? In addition, in the field of technology, India has largely relied so far on imported technology and lacks the capability to develop "high tech products based on indigenous genetic resources on our own" (Gadgil, 1995:317) .

This tension may also be seen to translate into an uncomfortable anomaly in the positions taken by environmental NGOs. On the one hand, they must support national sovereignty in the face of external threats, whether from other states or from transnational corporations. On the other hand, to the extent that the rights of local communities are not adequately safeguarded by national governments, they must also oppose the latter.

Typically, claims to rights—whether of nation-states or of local communities—invoke ideas of property and ownership, especially when physical resources are in question. But is a right to property the best and most efficacious form of expressing such rights-claims? What are the implications of appealing to the concept of property and seeking to alter only its rights-holders? The political position that repeatedly emerges from environmental activism is that nation-states should have sovereign rights of ownership and control over their biological resources; that local communities should be recognized as the owners and creators of indigenous knowledge systems and indigenous technology; and that future generations are potential owners, as the inheritors of this legacy. All these claims strongly echo the imagery of property. The concept of property recalls not only the idea of commoditization, but also an attendant notion of inequity and lack of participation.

This paper, therefore, questions the appropriateness of couching such claims in the vocabulary of property rights and suggests a shift in the way in which rights are explicated in this area. Drawing upon the interpretation of rights as self-ownership (as evidenced in much liberal political theory, especially the influential work of Robert Nozick, 1974) versus rights as self-government (Ingram, 1994), this paper suggests that a shift away from the self-ownership view and towards the view of rights as self-government is appropriate. The self-ownership view, heavily imbricated as it is in the language of property, suggests control and command through ownership and is, eventually, the language of power and domination. Interpreting rights in terms of self-government, on the contrary, yields a more egalitarian, participatory, and democratic perspective—one which is infinitely more suitable for an ecologically sensitive approach. The inequities and domination which are implicit in the interpretation of rights as self-ownership (and, thereby, linked inextricably to the phenomenon of property) are at least philosophically redressed and a more radical ecological agenda rendered possible.✧

Selected Reading

Chatterjee, Pratap and Finger, Matthias, 1994. The Earth brokers: power, politics and world development. Earthscan Publications, London.

Feinberg, Joel, 1980. Rights, justice and the bounds of liberty. Princeton University Press, Princeton, NJ.

Gadgil, Madhav, 1995. Patenting life. In: Artha Vijnana, Vol. 37, No. 4, December.

Ingram, Attracta, 1994. A political theory of rights. Clarendon Press, Oxford.

Johnson, Lawrence E., 1993. A Morally deep world: an essay on moral significance and environmental ethics. Cambridge University Press, NY.

Naess, Arne, 1991. Deep ecology. In: Dobson, Andrew (ed.), The green reader. Andre Deutsch, London.

Nozick, Robert, 1974. Anarchy, state and utopia. Basic Books, New York.

O'Neill, John, 1993. Ecology, policy and politics. Routledge, London.

Pearce, David and Morgan, Dominic, 1994. The economic value of biodiversity. Earthscan Publications, London.

Sahai, Suman, 1995. Biotechnology: new global money-spinner. In: Economic and Political Weekly. Volume XXX, No. 46, November 18.

Shiva, Vandana, 1996. Agricultural biodiversity, intellectual property rights and farmers rights. In: Economic and Political Weekly. Vol. XXXI, No. 25, June 22.

Singer, Peter, 1993. Practical ethics. Cambridge University Press, Cambridge.

Warren, D. Michael, Slikkerveer, L. Jan, and Brokensha, David (eds.), 1995. The cultural dimension of development: indigenous knowledge systems. Intermediate Technology Publications, London.

Plant Breeders and Farmers in the New Intellectual Property Regime: Conflict of Interests?

Biswajit Dhar[1] and C. Niranjan Rao[2]
[1]*Research and Information System for the Non-Aligned and Other Developing Countries, Indian Habitat Centre, New Delhi*
[2]*Indian Council for Research in International Relations, India Habitat Centre, New Delhi, India*

Introduction

The issue of intellectual property protection (IPP) has assumed significance in most developing countries, particularly in light of their commitments to bring about far reaching changes in the norms and standards of protection of intellectual property. These commitments form part of the Agreement on Trade Related Aspects of Intellectual Property Rights (TRIPs), formalized at the end of the Uruguay Round of Multilateral Trade Negotiations. As a result, developing countries have undertaken to amend their existing regimes for the protection of intellectual property and adopt ones which are quite akin to those prevailing in the industrialized countries. The harmonized system of intellectual property protection (IPP) that would thus come into existence has two major departures from the hitherto existing international system—

- the scope of protection afforded by the IPP regime would be considerably wider
- the rights of the owners of intellectual property would be stronger

In the former instance, the most significant is the requirement to extend IPP to agriculture by protecting improved varieties of plants.

The Agreement on TRIPs requires countries to introduce either patent or an effective sui generis system to cover improvements in plant varieties (emphasis added). This requirement, as we shall discuss, means that countries would have to introduce the system of Plant Breeders Rights (PBRs) which is internationally recognized through the UPOV Convention (International Convention for the Protection of New Varieties of Plants). However, this manner of protection provides explicit recognition only to the rights of commercial plant breeders engaged in developing new plant varieties. For several developing countries like India where agricultural systems have been based almost exclusively on farming communities, this form of IPP raises several key issues from the viewpoint of the farmers.

This paper discusses two dimensions of impact that the new regime of IPP would, in our view, have on farmers in most developing countries. In the first place, the system of PBRs would impinge on the traditional agricultural practices of the farmers. Secondly, and more importantly, the new Intellectual Property (IP) regime discriminates against the farming communities by recognizing only the contribution made by the modern plant breeders while remaining silent about the contribution made by the farmers in both development of new varieties of plants as well as conservation of biodiversity.[1] Farmers and plant breeders, particularly those in the public sector, have in fact formed symbiotic relationship in the process of bringing about improvements in plant varieties, but this dimension of agricultural innovation has been given a short shrift in the new IP regime.

The two dimensions of impact on farmers would be discussed in the sections that follow. In the first, the nature of the framework for plant variety protection would be elucidated. This discussion would highlight the UPOV system of Plant Breeders Rights (PBRs) and the rights that the plant breeders are allowed under this system, particularly in relation to the users of the plant varieties in developing countries; i.e., farmers. The second section would discuss briefly the part that the farmers have been playing in the development of agricultural technology. This discussion highlights the point that the farmer has not only been playing an important role in the process of technology generation through informal innovations. His role in developing a viable agricultural system that can fulfil the needs of the multitude of poor farmers is extremely crucial. But instead of recognizing their role in the process of agricultural change, the IPR regime excludes them.

An Evaluation of the System of Plant Varieties Protection

The need to extend IPP to agriculture, we had stated at the outset, arises out of the commitments made by the countries laid down by the Agreement on TRIPs. One of the main aims of the Agreement is to ensure that "adequate standards and principles concerning the availability, scope, and use of intellectual property rights" are provided by all countries. This objective was laid out in furtherance to the global efforts at ensuring that norms and standards of IPP are harmonized across countries. Towards this end, initiatives were taken in the past by the World Intellectual Property Organization (WIPO), for evolving a harmonized system of protection, particularly in the area of patents. The inter se differences in the framework of patent protection as between countries was thus sought to be eliminated. This is an important consideration that needs to be kept in view when interpretations of the Agreement on TRIPs and its applicability to the area of plant varieties protection are made. What makes the above reference to the WIPO efforts at harmonization of patent laws more pertinent in case of plant varieties protection is that the Agreement on TRIPs proposes the same framework for both these forms of IPRs and are, hence, proposed to be dealt with in a similar manner.

The norms for the protection of plant varieties are specified in the Agreement on TRIPs in Article 27.2(b), and although at first sight the scope of the relevant provision appears open to interpretation, in reality, however, it is quite well defined. Article 27.2(b) provides that "members shall provide for the protection of plant varieties either by patents or effective *sui generis* system or by any combination thereof" (emphasis added). The qualification made of the *sui generis* system that is to be introduced as per the requirements of this Article holds the key to the adoption of the framework for the system. The sui generis system that countries are required to introduce have to be "effective" and the effectiveness of the system would be determined in the WTO. Globally, the only "effective" *sui generis* system existing for the protection of plant varieties at present is the PBRs defined by the UPOV Convention that has been in existence since 1961. Considering its wide acceptance among the developed countries which have pushed for extending of IPP to agriculture in all countries, the UPOV framework can be considered to be in conformity with the requirements of Article 27.2(b).

In this context what needs further consideration is that the UPOV has amended the framework of PBRs in 1991.[2] The amended UPOV of 1991 (UPOV 91) was ratified by 17 countries by 1995[3] (the United States was the first to do so) and is increasingly being considered as the framework towards which all the national systems must gravitate. It has also been argued that UPOV 91 takes into consideration the recent developments in biotechnology and hence all the state-of-the art technologies in plant breeding can be covered by this form of PBRs (Boeringer, 1992). It is therefore necessary to look at UPOV 91 to understand the nature of the system of plant varieties protection that countries would have to develop.

The main feature of UPOV 91 is that it seeks to enhance the rights of the modern plant breeders at the expense of the farmers. This has been done by: (i) strengthening the rights that the breeders can enjoy, and (ii) restricting the domain of operations of the traditional farmers, The above mentioned dimensions of breeders' rights would occur simultaneously through the singular move to extend the scope of their rights far beyond those available under UPOV of 1978 (UPOV 78). In the following discussion, we would try to argue that UPOV 91 is aimed at introducing a system of plant variety protection in which the modern plant breeders would have all pervasive control over agricultural practices. Not only would they be in a position to prevent reuse of the propagating material covered by the PBRs, the breeders could also exercise their control over the harvested material that use protected planting material and products made from such harvested material. This can be seen by comparing the rights available to the breeders in the new framework with those that were provided by the earlier framework (i.e., UPOV 78). The details are provided below.

The enhanced rights that the breeders can enjoy under UPOV 91 stems from the redefining of the scope of their rights. While under UPOV 78, breeders' rights had a rather limited scope, the amendment brought about in 1991 makes the rights all-encompassing. The former, in Article 5, provided that the breeder's rights extended to the following acts involving a protected plant variety: (a) production for purposes of commercial marketing, (b) offering for sale, and (c) marketing. These provisions indicated that authorization of the breeder was required if production of the planting material (primarily seeds) was undertaken only for commercial purposes. This, in other words, implied that the breeder's authorization was not necessary for reusing the planting material on the farm it was produced. The above interpretation gave rise to what has been commonly known as the "farmers' privilege" and it is taken to imply that farmers were permitted to reuse the propagating material from the previous year's harvest. Further, farmers were allowed to freely exchange seeds of the protected varieties with their farm neighbors. This was one of the prominent features of the US Plant Varieties Act of 1970 (Smith, 1996; Butler, 1996). These flexibilities available to farmers under UPOV 78 meant that their traditional practice of reusing farm-saved seeds and exchanging them with their neighbors could be carried on in an uninhibited manner.

It needs to be clarified here that the "farmers' privilege" under UPOV 78 is quite distinct from the concept of farmers' rights which was recognized by the international community while adopting the FAO Resolutions on Plant Genetic Resources, the process of which was initiated in 1983. This concept was formally put forth in three resolutions and was ratified by more than 160 countries in FAO Conferences in 1989 and 1991. A further enunciation of the contribution of the farmers in the "area of conservation and sustainable utilization of plant genetic resources, food, and sustainable agriculture" was provided in Agenda 21 adopted by the UNCED in 1992.[4]

An elaboration of the concept "farmers' rights" has been attempted in the draft Plant Varieties Act (PVA) proposed by the Indian Government. In the context of the present discussion it would be useful to consider the contours of the farmers' rights proposed by the draft PVA.[5] Clause 5 of the draft PVA provides for farmers' rights. According to this clause, the PVA "is to protect the rights of the developers of new varieties to stimulate investment in plant breeding and to generate competitiveness in the field of research and development both in the public and private sectors with the ultimate aim of facilitating access to newly developed varieties and maximizing agricultural production and productivity in the country...". The clause further states that "protection of farmers and researchers rights will strive to balance the need for stimulation and incentive to R&D with welfare of farmers".

The rights that the PVA proposes for communities and farmers are presented in the form of an incentive mechanism in Clause 22. Thus, the "recognition of the contribution made by rural communities with sustained perseverance in development, on-farm innovations, enrichment and conservation of plant genetic resources" is to come through "rewards and/or compensation to such communities or clusters...such that rural communities may have a stake in and continue their efforts at preservation and improvement of land-races...". In a similar vein, in recognition of their contribution in ensuring conservation, improvement and availability of plant genetic resources, the farmers are proposed to be given the rights to "secure full benefits and support in continuation of their contribution."

The farmer's rights are spelt out quite unambiguously in this clause in a further elaboration to the relevant clause. A farmer is provided "...additional rights to dispose of his farm produce as he chooses which includes his right to save, use, exchange, share and sell propagating material or seed obtained or descended from seed obtained of protected variety except sale of branded seed/propagating material...," although the last mentioned condition relating to branded seed/propagating material could imply a significant dilution of the rights in light of the fact that the large TNCs and their branded seeds are making inroads into the seeds market.

The concept of farmers' rights, although acknowledged by the FAO, suffers from the singular weakness in that there is no internationally accepted instrument through which it can be exercised. In the absence of such an instrument to put into operation the farmers' rights, the breeders' rights as defined by the PBRs, have come to be accepted as comprising the *sui generis* system of protection available for plant varieties worldwide. Thus, not only does UPOV 91 not give any recognition to the farmers' rights, it has proposed stronger rights for the breeders.

The scope of the breeders' rights is provided by UPOV 91 in Article 14. According to the provisions contained in this Article, the breeders could extend their rights, not merely to the propagating material, but to the harvested material and the products made of the harvested material, in case the production process uses varieties protected by PBRs. The breeder rights are defined as including the following acts: (i) production or reproduction (multiplication), (ii) conditioning for the purposes of propagation, (iii) offering for sale, (iv) selling or other marketing, (v) exporting, (vi) importing, (vii) stocking for any of the purposes referred above.

The rights of the breeder, mentioned above, has been extended to cover all acts pertaining to production and reproduction of the propagating material on which his rights have been established (Article 14(1) of UPOV 1991). The scope of protection thus leaves virtually no possibility of farmers reusing seeds without authorization of the breeder as under UPOV 78. Nominal scope for exception to breeders' rights have, however, been provided under 15.2 as under "... each Contracting Party (to UPOV 91) may, within reasonable limits and *subject to the safeguarding of the legitimate interests of the breeder*, restrict the breeder's right in relation to any variety in order to permit farmers to use for propagating purposes, on their own holdings, the product of the harvest which they have obtained by planting, on their own holdings, the protected variety...". (emphasis added).

This is in sharp contrast to the earlier system under which the farmers were allowed to reuse and also exchange the protected material with their farm neighbors without paying any royalty to the commercial breeders, as mentioned earlier. The new provisions thus allow the farmers to reuse the protected material only if the "legitimate interests of the breeder" are taken care of, the "legitimate interests" being the royalty that the breeder should be paid for reusing the seeds.

The recent amendment of the PBR system brought about by the US Congress after ratifying UPOV 91 pointedly indicates the restrictions that farmers could face in the new system of PBRs. Through this new legislation, the US Congress has put limits on the scope of "farmer's exemption" under the US Plant Variety Protection Act (the U.S. equivalent of PBRs). According to this legislation, farmers would be allowed to replant the seeds on their own farm but would be restricted from selling them for reproductive purposes to their farm neighbors without having to pay royalties or ask permission for the same (Hamilton, 1996).

The breeders' control over production system that uses protected varieties of plants has been strengthened by extending the scope of PBRs to include harvested material and products made directly from the harvested material. Article 14(2) of UPOV 91 provides that breeders' rights could extend to commercial acts "in respect of harvested material, including entire plants and parts of plants, obtained through unauthorized use of propagating material of protection variety....". Article 14(3) extends the breeders' rights to products "directly made from harvested material of the protected variety....". In both these cases, commercialization of the products of the harvest by anyone other than the breeder would not be permitted until the breeder has exercised his options to commercialize the products arising out of the protected material. However, the rights of the breeder are seen to be exhausted at the first act of commercialization, involving the propagating material or harvested material including entire plants or parts of plants or products made directly from harvested material (Article 16).

Article 14(5) of UPOV 91, which provides for the inclusion of "essentially derived varieties" of the protected varieties within the scope of the rights of the breeder, constitutes a major element in this latest amendment to the UPOV. This provision expressly seeks to strengthen the rights of the breeder by bringing within the purview of protection "essentially derived and certain other varieties" of the protected varieties.

In introducing the above-mentioned provision, the so-called "research exemption" available under UPOV 78, which allowed breeders to freely use other breeder's protected varieties for research purposes and for breeding new varieties, was sought to be excluded. It was argued that the benefits that a breeder could secure was limited since "research exemption" allowed creation of a new variety of plant by using protected varieties without authorization of the original breeder[6]. Article 14(5) was accordingly introduced to ensure that no new variety can be produced by using the protected varieties by any means. An "essentially derived variety" has been defined by UPOV 91 thus:

(i) it is predominantly derived from the initial variety, or from a variety that is itself predominantly derived from the initial variety, while retaining the expression of the essential characteristics that result from the genotype or combination of genotypes of the initial variety

(ii) it is clearly distinguishable from the initial variety

(iii) except for the differences which result from the act of derivation, it conforms to the initial variety in the expression of the essential characteristics that result from the genotype or combination of genotypes of the initial variety.

This article further[7] provides a nonexhaustive list of examples of acts that may result in the essential derivation, including the selection of a natural or induced mutant, or of a somaclonal variant, the selection of a variant individual from plants of an initial variety, back-crossing or transformation by genetic engineering. This indicates that all acts of breeding, from the most conventional to the one involving use of modern techniques, would be taken into consideration while determining whether or not a new variety is "essentially derived."

The strengthening of the rights of the breeders took place despite the fact that several countries (more prominently Japan and Canada) had raised the issue in the Diplomatic Conference leading up to the adoption of the revised UPOV that identification of "essentially derived varieties" was a controversial issue. These countries had emphasized that prior to the assigning of rights for an essentially derived variety in line with the provisions of the article providing such protection, effective guidelines must be laid down for identifying such varieties. In this context it needs to be mentioned that UPOV 91 took note of this observation through a decision to evolve the guidelines.

Introducing the concept of essentially derived varieties raises several other contentious issues. There is a view that determination of the such derived varieties would not be made by an examining office as a part of the grant of PBRs but between plant breeders either through a mutually arrived agreement or through litigation.[8] This implies that this critical issue would be settled by the relative strengths of the parties involved, an eventuality that may not favor developing countries like India in which plant breeding, both in the formal and the informal sectors, has a long history.[9]

The above discussion shows quite clearly that UPOV 91 would establish unambiguously the dominance of the modern plant breeders in the agricultural sector, as a reward for their contributions to the advancement in technology. This reward system, which forms the basis of the PBRs, is structured in a manner that excludes the farmers. The primacy accorded to the breeder through the PBR system affects the traditional farming communities in two principal ways. In the first instance, the contribution made by the farmers in the development of technology does not find any recognition. Secondly, the move towards making the innovation activity an exclusive monopoly of the modern breeders forecloses any future farmer participation in research. The latter, in particular, has been indicated as a possible route for ensuring that technologies which can be adopted by the resource poor farmers are developed. These two dimensions of farmer-contribution to innovation are discussed below.

Farmers as Innovators

On-farm innovations by farmers has taken place as a continuous process throughout the period of settled agriculture. However, the history of recognizing this important contribution of the farming communities does not extend very far back in time. But in recent years, a vast body of literature has been generated on this issue, particularly in response to the global initiatives at preserving and conserving of biodiversity and the role of the indigenous people in these conservation activities. Given such a wealth of available documentation, we would attempt to keep the discussion down to the bare essentials.[10]

The nature of farmer-innovation is succinctly presented by Bigs and Clay (Bigs and Clay, 1981) as under: "Farmers select by identifying and using plants of economic importance, continually retaining and reusing seed, and propagating material with preferred characteristics, such characteristics relate to a whole series of attributes which the farmer comes to associate with performance in terms of yield, robustness, and suitability for time bounded production." This nature of purposive selection that the farmer has been engaged in, has been recognized by the authors as one source of innovation which is implemented by informal experimentation.

The above mentioned description of the farmer-innovation process wherein the farmers are seen to be adopting a clearly defined criteria to identify the improved varieties they have been able to develop, has a certain resemblance with that followed by the modern plant breeders. The latter rely on the three-fold criteria of distinctiveness, uniformity, and stability (DUS) of the plant varieties they develop in order to lay claim on the breeders' rights. But even when the farmer-innovation process has been seen to have an objective basis in so far as the criteria for identifying the improvements are well laid out, no recognition is accorded to this system of innovation akin to that available to the modern system of plant breeding. The recognition that the modern plant breeders receive is facilitated by the extensive codification of their knowledge, while the lack of such codification lies at the heart of the relative neglect of the contribution that the farmers have made in the development of agricultural technology over generations.

One of the impediments in attempting codification of the farmers knowledge is the unique failure of the farmer-innovation process. Unlike the modern breeder who conducts his research under a controlled environment, the farmer uses the available environmental conditions to conduct his experiments. Furthermore, the use of the environmental conditions based on a detailed knowledge of the environment in which they practice their farming. It has been indicated that natural selections occurs "by the action of environmental stress on inherent variation caused by gene recombination and mutation" (Bigs and Clay, 1981). This process of natural selection is seen to occur not only in the plants that are useful to the human kind but also in case of other organisms which directly affect the process of agricultural production. Included in the latter category are weeds, pests, and other microorganisms. The farmers reliance on natural selection has thus resulted in a continuous and evolving process of change that agriculture has been subjected to and has not brought about adaption within a specific or given environment. The farmer innovation process is succinctly summed by Bigs and Clay thus: "The farmer is not moving iteratively towards some optimal point, but is only able to stay in dynamic equilibrium with his environment by continuous innovation." The "continuous innovation" that the authors refer to would be supported below by citing a few disparate cases of farmer innovations.

Quite apart from the past contributions that farmers have made in the generation of technologies, an important future role for the farmers in the process of agricultural innovation has been emphasized in several studies. These studies have argued that the formal system of research has a distinct rich farmer bias and as a result the vast multitude of poor farmers in most of the developing countries are unable to benefit from the so-called advances in farm technology. This bias, the studies have indicated, can be rectified by involving farmers in research "at all levels and stages and sharing credit for results" (Harwood, 1979).

Chambers and Jiggins provide an informed overview of the early studies on the issue. Most of these studies quoted by the authors have tried to argue that given need to focus agricultural research towards the requirements of the poor farmers, the research priorities should be allowed to be set by the poor farmers themselves. Agricultural research, the authors argue by quoting an earlier study, should begin "with a systematic process of scientists learning from, and understanding, RPF (resource poor farmer) families, their resources, needs, and problems. The main locus of research and learning is the resource-poor farmer, rather than the research laboratory." This approach, wherein the RPF families themselves identify priority research issues, is, according to Chambers and Jiggins, "based on respect for and confidence in the ability of RPF families to tell scientists their understanding of the problems they face and to identify how the formal research system can help them."

The idea of a "turnaround" in agricultural research with the RPF families laying down the priorities as proposed by Chambers and Jiggins among others is, however, not shared by some of the other commentators on farmer participatory research. Tripp, for instance, maintains that farmer participation should be central to adaptive agricultural research while holding the view that reorientation of research based on the farmers' knowledge system may not be in order. Among the arguments presented in support of this view is that resource poor farmers who form the core of the idea of reorienting agricultural research, would not be in a position to perform the role they are seen to be playing, primarily because of two factors. One, the poorest are unlikely to develop institutions and two, as Biggs and Clay have observed, informal R&D involves certain costs which the poor farmers may not be able to bear. While some have argued that the real benefits would occur only with a sizeable farmer participation (Chambers and Jiggins, and Biggs and Clay), others like Tripp have tried to draw the bottom line. Interest in farmer participation, according to the latter view, "has served to draw attention to several key issues that agricultural research has yet to solve" (Tripp, 1989).

Conclusions

The issue of extending intellectual property protection (IPP) to agriculture has assumed significance, particularly in developing countries, in the context of the extensive changes which could take place in the peasant-dominated agricultural systems in these countries. The importance of the changes, as we have argued, lies in the fact that countries are expected to adopt a framework of IPP having a strong bias in favor of the modern plant breeders. This form of IPP has been proposed in recognition of the contribution that the breeders are seen to be making in the development of agricultural technology. The reward system through the grant of IPRs thus proposed, discriminates against the traditional farmers in developing countries. Not only would the latter encounter problems in continuing with their traditional practice of retaining a part of their harvest as seeds, the breeders can also exercise their rights over the harvested material. By according such sweeping rights to the breeders, the proposed system of IPP almost completely excludes the farmers from the agricultural system.

We have tried to argue against this exclusion that the farmer faces in the proposed regime. Our view is that the farmer needs to be recognized because of two compelling factors. One, the farming communities have played a crucial role in the process of technology generation. And, two, a farmer-centered innovation process has been seen as being a more appropriate system for meeting the needs of the resource-poor farmers. Thus, if viability of agricultural systems in the farmer-dominated agricultural systems is to be ensured, it is imperative to look beyond the confines of the narrowly defined regime of plant breeders' rights.✧

†*The views expressed here are the authors' own and do not necessarily reflect those of the organizations to which they belong.*

Footnotes

1. *See for example the discussion in Wood, 1988.*
2. *This amendment was the third following the two earlier amendments in 1972 and 1978.*
3. *FAO, 1995, Appendix 2.*
4. *Chapter 14.*
5. *Our commentary on the PVA is based on an earlier draft of the proposed legislation. This would not affect the present discussion since the recent draft has features similar to the earlier draft, as is evident from the outline of the former that we have access to.*
6. *Articles 5(3) allowed use of a protected variety as an initial source of variation for the purposes of creating other varieties.*
7. *Article 14(5)(c).*
8. *Greengrass, 1993.*
9. *For a brief historical account of agricultural innovations in ancient India, see ICAR, 1964.*
10. *See for example IDRC, 1994.*

Selected Reading

Biggs, Stephen and Clay, E.J., 1981. Sources of innovation in agricultural technology, World Development, vol. 9. no. 4.

Boeringer, Dirk, 1992. Developments in biotechnology and the 1991 Act of the UPOV Convention in UPOV, Seminar on the Nature of and the Rationale for the Protection of Plant Varieties under the UPOV Convention, held in Tsukuba, Japan, November 1991.

Butler, L.J., 1996. Plant Breeders' Rights in the US: update of a 1983 study. In: van Wijk, J. and W. Jaffé (eds.), Intellectual property rights and agriculture on developing countries, University of Amsterdam.

Chambers, R. and Jiggings, J., 1986. Agricultural research for resource poor farmers: a parsimonious paradigm. IDS discussion paper, # 220, August.

FAO, 1989. Interpretation of the international undertaking on plant genetic resources. c 89/24, Rome.

FAO, 1993. Plant breeders' rights: India: terminal statement prepared for the government of India by the Food and Agriculture Organization of the United Nations. AG:TCP/IND/0052, Rome.

FAO, 1994. Commission on Genetic Resources for Food and Agriculture, revision of the international undertaking, Issues for consideration in Stage II: access to plant genetic resources and farmers' rights, CPGR/94/WG9/4.

FAO, 1995. Recent international developments of relevance to the draft code of conduct for plant biotechnology, CPGR-6/95/15.

Greengrass, Barry, 1991. The 1991 Act of the UPOV Convention", in UPOV, seminar on the nature and rationale of plant varieties under the UPOV Convention, Beijing, China, September, 1993.

Hamilton, Neil D.,1996. Possible effects of recent developments in plant-related intellectual property rights in the US. In: J. van Wijk and W. Jaffé (eds.), Intellectual property rights and agriculture on developing countries, University of Amsterdam.

Harwood, Richard R., 1979. Small farm development: understanding and improving farming systems in humid tropics, quoted by Chambers, R. and Jiggings, J., 1986.

ICAR,1964. Agriculture in ancient India, New Delhi.

IDRC, 1994. People, plants and patents: the impact of intellectual property on trade, plant biodiversity and rural society. Report prepared by the Crucible Group, Ottawa.

Maxwell, Simon, 1986. Farming systems research: hitting a moving target." World Development, vol. 14, no. 1.

Shand, Hope, 1994. US Congress restricts farmers' rights". Seedling, October.

Smith, Stephen, 1996. Farmers privilege, breeders; exemption and essentially derived varieties concept: status report on current development. In: J. Wijk and W. Jaffe (eds.), Intellectual property rights and agriculture on developing countries, University of Amsterdam.

Tripp, Robert, 1989. Farmer participation in agricultural research: new direction for old problems. IDS discussion paper # 256, February.

UN, 1992. Agenda 21, New York.

UPOV, 1991. Seminar on the nature and rationale of plant varieties under the UPOV Convention, Budapest, Hungary, September, 1993.

UPOV, 1992. Records of the Diplomatic Conference for the Revision of International Convention for the Protection of New Varieties of Plants, Geneva.

Wood, David, 1988. Crop germplasm: common heritage or farmers' heritage? In: J. Kloppenberg (ed.), Seeds and sovereignty: the use and control of plant genetic resources. Duke University Press, Durham.

Aspects of Biosafety in the Conservation of Biological Diversity

K. P. S. Chauhan
Ministry of Environment and Forests, Government of India, Paryavaran Bhawan, New Delhi, India

Introduction

The rapid unfolding of vast opportunities for applying the tools and techniques of biotechnology in the areas of agriculture, health, industry, and environmental pollution have aroused tremendous and wide ranging expectations. These technologies have the potential to provide more abundant and nutritious food, new medicines including therapy for genetically induced diseases, new environmentally friendly products, and the means to clean up industrial pollution of water and soil. These technologies are also being applied in the assessment, monitoring, management, and sustainable utilization of biological diversity.

As is increasingly being realized, it is necessary to create the supporting infrastructure and a congenial environment to capitalize on the potential of the new biological technologies (Brenner, 1995). It has also been acknowledged that the release of living modified organisms (LMOs) in a contained or open environment could pose risks that would have various direct and indirect impacts. If the benefits of biotechnology are to be optimized without affecting the environment, effective biosafety regulations must be developed based on sound scientific principles (Persley et al., 1992; Walsh, 1993; Krattiger and Lesser, 1994).

Dealing briefly with the potential of biotechnology and its impact on the environment (in particular, on biological diversity), this paper traces the various developments that have taken place in developing biosafety measures at the national and international levels, including the finalization of the International Technical Guidelines for Safety in Biotechnology (UNEP, 1995). In the light of the decision taken by the Conference of Parties of the Convention on Biological

Diversity (CBD) at its second meeting held in Indonesia in 1995, the paper also highlights the basic elements which are being considered for inclusion in the proposed "Protocol on Biosafety" under the aegis of the CBD. It also discusses which elements are to be negotiated by the contracting parties.

Impacts of Biotechnology on Biological Diversity

Assessment, Monitoring, Managing, and Sustainable Use

Biotechnology provides a range of tools and methods for assessment, monitoring, and managing biological diversity, such as clarifying taxonomic and evolutionary relationships among groups of organisms and assessing the effects of ecosystem disturbance on components of biological diversity and biological processes. Besides assessment, monitoring, and managing, the tools of biotechnology could also be utilized for *in situ* conservation; i.e., assessment of optimal or minimal population size, and *ex situ* conservation; i.e., enhancing the quality of characteristics and efficiency, through compact storage of DNA libraries and sequence databases (Appels et al., 1995). The details are given in Figure 1. The tools currently employed for the sustainable use of biological diversity are in the area of breeding, genetic engineering, the development of novel genes and gene products, and environmental remediations. Some areas of application of biotechnology in utilization of biological diversity, as identified by Montagu et al. (1995), are given in Table 1.

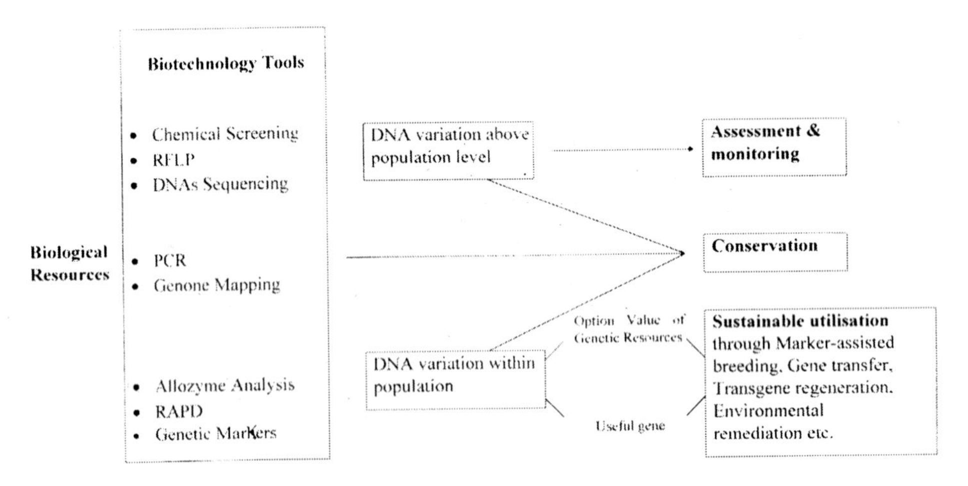

Figure 1. Biotechnology applications in assessment, monitoring, conservation, and utilization of biological diversity. (Source: Appels et al., 1995.)

Impacts of Biotechnology

The application of biotechnology in different areas using the components of biological diversity can have impacts which may be both direct and indirect. These impacts mainly arise from the release of living modified organisms and their interactions with the environment.

Direct Impacts

The introduction of any LMOs in a biological community can have the following various undesirable impacts (Tzotzos et al., 1995):

- Displacement or destruction of indigenous/endangered or endemic species
- Exposure of species to new pathogenic or toxic agents
- Pollution of the gene pool
- Loss of species diversity
- Disruption of energy and nutrient cycling

These impacts are largely ecological and evolutionary and can be assessed scientifically and tested through simulated conditions. As part of an overall risk assessment strategy, the main problems arising from direct impact could be dealt with, including the processes of introgression, weediness, pathogenicity, and altered nutrient cycling. However, there is an inadequate understanding of the possible direct impact of LMOs on soil micro flora and fauna (Angle, 1994 and Morra, 1994), and the potential of virus-resistant plants on the host range of some viruses (Rissler and Mellon, 1993).

Indirect Impacts

The possible number and types of indirect effects of biotechnology could be immense. These effects are mostly socioeconomic in nature and can be of major importance, particularly to low or middle-income developing countries where people are dependent on biological resources for subsistence. Indirect impacts may be secondary or tertiary effects. Tzotzos et al. (1995) has listed the following indirect impacts:

- Pressure on natural habitats because of the increasing value of genetic resources
- Lack of immediately perceivable incentives for conservation
- Moral/ethical problems of ownership of genetic resources and benefit-sharing
- Increase in agricultural productivity
- Replacement of traditional landraces
- Decline or loss of opportunities loss for disadvantaged groups in areas of marginal production

Public Perceptions

The public debate on the applications of biotechnology has been marked by apprehensions of two kinds. The first is that there may be adverse impacts on the environment and human health. The second is the fear that control of the new technologies could give some nations or groups the power to use this in unfair ways.

The public perception of biotechnology as unpredictable and dangerous was highlighted by Perlas (1993) through examples such as the creation of a "Super Aids Virus"; super pigs and cows; reporting of serious ailments using a genetically altered version of L-tryptophan and insulin; the use of biotechnologically developed bovine growth hormone (BGH); stunting in corn and other crops due to the presence of Clavabacter xyli (a vector used to transfer the BT endotoxin gene); the requirement of six times more pesticides for Uniliver cloned oil palms, novel mutant carps, catfish, trout and salmon polluting native species; and illegal trials on pseudorabies in Argentina. This short list of unpredictable, and negative impacts of biotechnologies clearly shows that there will always be the adverse impact of biotechnologically developed products on human health and the environment. The views of Holmes (1993), Williamson (1991) and Ellstrand and Hoffman (1990) further support these perceptions.

Such hazards had been recognized by scientists even when the advances in biotechnology were in their infancy. Berg et al. (1974) voiced their concern about the potential biological hazards with respect to r-DNA experimentations. This led to the finalization of "The Guidelines for Research Involving r-DNA" by the

United States (U.S.) Government. By the mid-1980s, however, the context of biotechnology had shifted from research to commerce. Public perception and policy issues revolved particularly around biotechnological applications in nonmedical areas such as agriculture, bioremediation, and industrial aspects, including the safety of and the risks associated with these applications. An intense debate ensued between molecular biologists and ecologists on controversial issues regarding risk assessment related to the release of new biotechnology products into the environment (Krimsky, 1991). The debate was also joined by prominent environmental groups, politicians, corporations, and trade unions. This led to the formulation of a "Coordinated Framework for Regulation of Biotechnology" by the U.S. Subsequently, biosafety guidelines were developed by the OECD and European Union (EU) and its member countries. Several surveys conducted to analyze public perceptions revealed that they were very similar, in spite of geographical situations. However, there were also considerable differences between EU countries [(OTA, 1987); Hoban and Kendall (1992); and Marlier (1992)].

Table 1. Areas of application of biotechnology in biodiversity conservation

Using technologies as sources of—

- Proteins and peptides
- Lipids and fatty acids
- Carbohydrates
- Secondary metabolites for pharmaceuticals, food additives, and biopesticides

Genetic engineering, breeding and *in vitro* culture systems can be used to enhance agronomic performance.

Improving environmental conditions through—

- Identification of soil microorganisms and determination of best combinations for soil rehabilitation
- Use of plants to mitigate heavy metal pollution
- Engineering key genes in bacteria for pollutant degradation
- Improving plant microbe symbiotic systems for waste water treatments
- Production of biosurfactants, bioplastics, and other biodegradable products

Enhancing the efficiency of microorganisms in industrial processes such as—

- Microbial-enhanced secondary recovery of oil from reservoirs
- Bioleaching: microbiological extraction of metals from ? low-grade ores
- Production of industrial enzymes
- Production of endogenous products; e.g., antibiotics, alcohol, and organic acids

(Source: Montagu et al., 1995)

Table 2. Status of adoption biosafety regulations in different countries

Industrialized Countries	Developing Countries
Australia	Argentina
Austria	Brazil
Belgium	Chile*
Canada	China*
Denmark	Costa Rica*
Finland	Cuba*
France	Egypt*
Germany	Hungary**
Greece	India*
Ireland	Indonesia*,**
Israel	Kenya*,**
Italy	Malaysia*,**
Japan	Mexico
Luxembourg	Nigeria*
New Zealand	Philippines*
Norway	Russia**
Portugal	Thailand*
South Africa	Zimbabwe*,**
Spain	—
Sweden	—
Switzerland	—
The Netherland	—
United Kingdom	—
United States	—

** Lower-middle to low income economy.*
***Currently drafting regulations.*
(Based on Virgin et al., 1995)

Current Legislative and Regulatory Mechanisms for Biosafety

The unpredictability and hazards of using biotechnology and its possible misuse, as perceived by the public, could be countered through setting up an adequate system of safeguards and regulations. The current global status of regulations relating to biosafety has been extensively analyzed by Virgin et al. (1995). As per their analysis, 24 countries with high to high-middle income economies have laws and regulations in place, taking into account the specific concerns arising from new recombinant techniques. The EU countries have instituted new laws, which are similar in scope, requirements, and impacts.

In developing countries the situation is significantly different. Latin America, Argentina, Brazil, Mexico, Chile, Costa Rica, and Cuba have regulatory mechanisms in place. The African continent is represented only by South Africa and Egypt. Kenya, Zimbabwe, and Nigeria are at various stages of drafting regulations and are likely to finalize them very soon. In Eastern Europe, Hungary has an ad hoc review process, and Russia has submitted a biosafety law for official approval. Of the developing countries in Asia, only India, China, Thailand, and the Philippines have guidelines. Malaysia is preparing new legislation, and Indonesia is in the process of drafting some (Table 2).

It may be stated that the implementation of these biosafety regulations varies from developed countries to developing countries, from being very rigidly effective to noneffective because of the lack of a well-defined institutional structure. It is also pertinent that the guidelines evolved by most developing countries are very similar in scope and requirements and have been adopted as a part of their National Environment Acts. India has ratified Rules on Biosafety (1989) and revised Guidelines (1990) under the Environment (Protection) Act (1986) which governs the risk assessment and biosafety of hazardous microorganisms and/or genetically modified organisms or cells, with an appropriate institutional mechanism. However, these provisions are inadequate in respect of modalities/protocols for access to and transfer of biotechnology on "Mutually Agreed Terms," procedures for "Advanced Informed Agreements," and procedures for risk assessment and management (Chauhan, 1996).

Virgin et al. (1995) also studied the rate of adoption of guidelines by different countries. Analysis revealed that by 1997, 67 percent of countries with high to upper-middle income economies and 12 percent of lower-middle to lower income countries would have regulatory procedures in place. It is expected that less than

30 percent of the lower-middle to lower income countries will have biosafety procedures by the year 2005. The accrual of benefits from biotechnological applications would be facilitated by the harmonized adoption of biosafety regulations.

Post-UNCED Scenario

The issues relating to safe application of biotechnology in relation to conservation and sustainable use of biological diversity found a prominent place in the negotiations towards finalizing the text of the CBD, which was adopted at the UNCED Earth Summit held at Rio de Janeiro in Brazil in June 1992. This historic summit also adopted other important documents, namely: the Rio Declaration, Agenda 21; Non-Legally Binding Principles on Forests; the Convention on Climate Change; and the Agreement to negotiate a Convention on Desertification.

Chapter 16 of Agenda 21, which deals with "Environmentally Sound Management of Biotechnology," specifically seeks to ensure safety in biotechnology development, application, exchange and transfer through international agreement on principles to be applied on risk assessment and management.

The CBD, signed by 171 countries including India, came into effect the 29th December 1993. Articles 8(g) and 19(3) and (4) of the CBD also address the issue of safety in biotechnology. Article 8(g) calls upon each contracting party to "establish or maintain means to regulate, manage or control the risks associated with the use and release of LMOs resulting from biotechnology which are likely to have adverse environmental impacts that could affect the conservation and sustainable use of biological diversity, taking also into account the risks to human health."

Keeping this in view, the Governing Council of the United Nations Environment Programme (UNEP) "affirmed the desirability of UNEP contributing to international efforts on biosafety, including the development of international guidelines" (decision 18/36 B) which may be used by the national governments, intergovernmental, private sector and other relevant organizations to provide safety in biotechnology.

The non-legally-binding "International Technical Guidelines for Safety in Biotechnology (ITGSB)" were finalized under the aegis of UNEP, through extensive discussions among most countries which had ratified the CBD. Six chapters of these guidelines spell out various elements dealing with general principles, assessment and management of risks, mechanisms at national and regional level for providing safety, information supply and exchange, and capacity building. These are based on common elements and principles derived from relevant national, regional and international instruments, regulations, and guidelines. The guidelines also address the human health and environmental safety of all types of applications of biotechnology, from research and development to commercialization of biotechnological products containing or consisting of LMOs. The adoption of these guidelines is expected to facilitate governments in taking appropriate actions towards developing mechanisms for evaluating biosafety, identifying measures to manage foreseeable risks, and also develop processes such as monitoring research and information exchange, all of which improve the safe application of biotechnology (UNEP, 1995). This would also facilitate the implementation of Article 8 (g) in those countries which have ratified the CBD.

The general principles followed in the guidelines are based on identifying any hazards, assessing and managing the risks (case by case and in a stepwise manner), and monitoring. The guidelines provide details relating to the assessment and management of risks, safety mechanisms at national and regional levels, safety mechanisms at the international level using information supply and exchange, and capacity building. It also highlights the stepwise approach and provides a framework on each aspect which is to be developed by the countries, either based on already existing mechanism or a totally new one. It is expected that implementation of these guidelines will provide an impetus to the uniform development of capacity at the international level for risk assessment and management associated with the release of LMOs arising from biotechnological applications and harmonization as emphasized by the UNIDO (UNIDO, 1990).

Ongoing Efforts for Developing Biosafety Protocol

Article 19(3) of the CBD dealing with handling of biotechnology and distribution of its benefits states that "The Parties shall consider the need for and modalities of a protocol setting out appropriate procedures, including, in particular, advance informed agreement, in the field of safe transfer, handling and use of any LMOs resulting from biotechnology that may have adverse effect on the conservation and sustainable use of biological diversity." Further, Article 19(4) states that "Each Contracting Party shall, directly or by requiring any natural or legal person under its jurisdiction providing the organisms referred to in paragraph 3 above, provide any available information about the use and safety regulations required by that Contracting Party in handling such organisms, as well as any available information on the potential adverse impact of the specific organisms concerned to the Contracting Party into which those organisms are to be introduced".

In the second meeting of the Conference of Parties to the CBD, held in Jakarta from November 6–17, 1995, it was decided to establish an Open-ended Ad Hoc Working Group under the Conference of Parties to develop a protocol on biosafety, taking into account these two paragraphs of Article 19, the principle enshrined in the Rio Declaration on environment and development and, in particular, the precautionary approach contained in Principle 15. The first meeting of this Working Group was held in Aarhus (Denmark) from 22–26 July 1996 and was attended by 92 countries, United Nations agencies, prominent NGOs, and representatives of the private sector.

The Working Group deliberated on various substantial matters. The salient points on which the group agreed are as follows:

"The development of the draft Protocol shall, as priority:

- Elaborate the key concepts and terms that are to be addressed in the process;
- Include consideration of the form and scope of advance informed agreement procedures; and
- Identify relevant categories of LMOs resulting from modern biotechnology."

Key Concepts and Terms To Be Addressed in the Process

With regard to elaboration of the key concepts and terms to be addressed in the protocol, discussion revealed that countries wanted various terms, which are of crucial importance, legally defined on the basis of mutual agreement. These terms include: LMOs, Transboundary Movement, Transfer, Safe Transfer, Competent Authority, Familiarity, Adverse Effects, Contained Use, Intended/Deliberate Use, Unintended Release, Focal Point, Risk Assessment, Risk Management, Modern Biotechnology, Advanced Informed Agreement, Prior Informed Consent, Minimum National Standards, Biosafety, Limited Field Trial, Handling of

LMOs, Use of LMOs, Centers of Origin, Centers of Genetic Diversity, Compensation, Open Environment, Open Field Trial and Accidents.

Form and Scope of Advance Informed Agreement (AIA) Procedures

The majority of countries believed that "AIA" procedures constituted a highly important part of a protocol dealing with transboundary movement of LMOs, taking into account the provisions of the Basel Convention and the operational guidelines and principles developed by the Forest Stewardship Council. The procedures should also include the notification of components and must deal with transboundary movements of LMOs and data relevant to safety and information contained therein. The other important aspect associated with this is capacity building, which must be an integral part of the notification in a manner that the AIA mechanism is workable and practical.

Relevant Categories of LMOs Resulting From Modern Biotechnologies

The priority activity regarding relevant categorization of LMOs resulting from modern biotechnologies is to establish a clear understanding of, and early agreement on, the classes of organisms under consideration in the negotiation process. An agreed categorization would help to establish which existing international agreements might be applicable to some categories of LMOs and relevant to developing a biosafety protocol. In addition, categorization, according to the degree of assessed potential risk to biological diversity, would appear relevant in considering AIA procedures. Apprehensions also have been raised that risk classification for LMOs would be unrealistic as biosafety risks associated with a given LMO would be different under different geographical, ecological, and climatic conditions.

The Working Group also exchanged views on the elements to be included in an international framework on biosafety. The suggested contents of the protocol on biosafety are given in Table 3. The countries negotiating the protocol agree on some of the elements (column 1 of Table 3). However, there are likely to be prolonged negotiations regarding other elements (column 2 of Table 3), because of differences in the perceptions of developing and developed countries. It is expected that the same spirit that brought about agreement on the complex issues contained in the Convention on Biological Diversity would also be evident in finalizing the protocol on biosafety.

Table 3. Suggested contents of a protocol on biosafety

Items included in all proposals	Items included in some but not all proposals
■ Title	■ Objectives
■ Preamble	■ Scope
■ Use of terms/definitions	■ Jurisdictional scope
■ Advanced informed agreements	■ Criteria for use of AIA or notification procedures
■ Information sharing	■ Notification procedures
■ Relationship with other international agreements	■ Considerations for risk assessment and risk management
■ Institutional framework for functioning of the protocol	■ Mechanism for risk assessment
■ Settlement of disputes	■ Mechanism for risk management
■ Amendment	■ Emergency procedures
■ Final clauses	■ Minimum national standards on biosafety
	■ Designation of competent authority and national focal point
	■ Capacity building
	■ Transport and packaging requirements for the transfer of LMOs
	■ Transboundary movement between Parties
	■ Illegal traffic
	■ Duty to re-import
	■ Public awareness
	■ Clearinghouse
	■ Mechanism for bilateral agreements
	■ Liability/liability and compensation
	■ Consultations on liability
	■ Monitoring and compliance
	■ Financial issues
	■ Socioeconomic considerations
	■ Review and adoption
	■ Signature
	■ Accession
	■ Right to vote
	■ Entry into force
	■ Reservation and declaration
	■ Withdrawal
	■ Depository
	■ Authentic texts
	■ Annexes
	■ Transboundary movement from a Party through States that are not Parties
	■ Technical information network

Summary and Conclusions

Rapid advances in evolving new tools and methods in biotechnologies and their use for tailoring new biological products have opened up new avenues for the assessment, monitoring, and sustainable use of biological diversity. The development of transgenics and their release into both contained and open environments have raised concerns relating to their adverse impact (direct or indirect) on human health and the environment. As a result, National Institutes of Health (NIH), European Union (EU), and Organization for Economic Cooperation and Development (OECD) guidelines were formulated to minimize the diverse impacts. After UNCED (1992), extensive debate between molecular biologists and ecologists continued. Agenda 21 and the Convention on Biological Diversity provided platforms for negotiating the non-legally-binding International Technical Guidelines for Safety in Biotechnology, finalized in 1995 under the aegis of UNEP. In the meanwhile, the 2nd Conference of Parties of the CBD decided to constitute an Open-ended Ad Hoc Working Group to develop a protocol on biosafety. The first meeting of this Group (July 1996) established the basis for the negotiation process which is to begin very soon. Both developed and developing countries will have to find common ground, particularly on the contentious issues such as Advanced Informed Agreements, notification procedures, mechanisms for risk assessment and management, emergency procedures, transboundary movements of LMOs, liability and compensation, monitoring and compliances, and socioeconomic considerations. Also related to these issues would be the financial requirements for successful implementation of the provisions of the protocol by all contracting parties.

It is the special responsibility of global molecular biologists and ecologists to facilitate the negotiation process, so that the protocol on biosafety can be finalized early. This would help in the suitable development of biotechnology to meet the objective of sustainable development, on the one hand, and the sustenance of biological diversity for future uses, on the other. This is one of the most difficult challenges for the entire global scientific community because the development of a protocol on biosafety will be a turning point of our modern times.✧

Selected Reading

Angels, J.S., 1994. Release of transgenic plants: biodiversity and population level considerations. Molecular Ecology 3:45–50.

Appels, R., Young, A.G., Tzotozos, G.T., Barlow, B.A. and Simoens, C., 1995. Biotechnology applications in biodiversity assessment and management. In: V.H. Heywood (ed.), Global biodiversity assessment. Cambridge University Press, Cambridge.

Berg, P., Baltimore, D., Boyer, H.W., Cohen, S.N., Davis, R.W., Hogness, D.S., Nathans, D., Roblin, R., Watson, J.D., Weissman, S. and Zinder, N.D., 1974. Potential biohazards of recombinant DNA molecules. Science 185,303.

Brenner, C., 1995. Technology transfer: public and private sector roles. In: J. Kommen, J.I. Cohen, and S.K. Lee, (eds.), Turning priorities into feasible programmes: proc. of a regional seminar on planning, priorities and policies for agricultural biotechnologies in Southeast Asia. Intermediary Biotechnology Service/ Nayang Technological University. The Hague/Singapore.

Chauhan, K. P. S., 1996. Legal requirements for fulfilling obligations under the Convention on Biological Diversity. India J. of Pl. Genet. Resources 9(1)(in press).

Ellstrand, N.C. and Hoffman, C. A., 1990. Hybridisation as an avenue of escape for engineered genes. Bioscience 40:438–442.

Hoban, T.J., and Kendall, P.A., 1992. Consumer attitudes about the use of biotechnology in agriculture and food production. U.S. Department of Agriculture, Raleigh, North Carolina.

Holmes, B., 1993. The perils of planting pesticides, 34 pp. New Scientist, 28 August.

Office of Technology Assessment, 1987. New developments in biotechnology—background paper: public perceptions of biotechnology, OTA–BP–45, United States Government Printing Office, Washington, DC.

Krattiger, A.F and Lesser W.H., 1994. Biosafety—an environmental impact assessment tool—and the role of the Convention on Biological Diversity, p. 353–366. In: A.F. Krattiger, W.H. Lesser, K.R. Miller, Y. St. Hill, and R. Senananayake, (eds.), Widening perspectives on biodiversity. IUCN, Gland, Switzerland and Int. Academy of the Env., Geneva, Switzerland.

Krimsky, S., 1991. Biotechnics and society—the rise of industrial genetics. Praeger Publisher, New York.

Mailier, E., 1992. Eurobarometer 35.1: opinions of Europeans on biotechnology in 1991. In: J. Durant, (ed.), Biotechnology in public: a review of recent research. Science Museum for the European Federation of Biotechnology, London.

Montagu, van M., Tiedje, J.M., Powell, D., Simoens, C., Tzotzos, G.T., and Barlow, B.A., 1995. Application of biotechnology for the utilisation of biodiversity. In: V.H. Heywood (ed.), Global biodiversity assessment. Cambridge University Press, Cambridge.

Morra, M. J., 1994. Assessing the impact of plant products on soil organisms. Molecular Ecology 3:53–55.

Perlas, N., 1993. When what could go wrong did go wrong, 8 pp. Third World Resurgence, Penang, Malaysia.

Persley, G. J., Giddings, L.V., and Juma, C., 1992. Biosafety: the safe application of biotechnology in agriculture and the environment. International Service for National Agricultural Research. The Hague.

Rissler, J., and Mellon, M., 1993. Perils amidst the promise: ecological risks of trasngenic crops in a global market. Union of Concerned Scientists, Cambridge, Mass.

Tzotzos, G.T., Lesser, W. H., Powell, P. J., amd Dale, P. J., 1995. Impacts of biotechnology on biodiversity. In: V.H. Heywood (ed.), Global biodiversity assessment. Cambridge University Press, Cambridge.

UNEP, 1995. International technical guidelines for safety in biotechnology. United Nations Environment Programme, Nairobi, Kenya.

UNIDO, 1990. An international approach to biotechnology safety, p. 43–47, UNIDO, Vienna.

Virgin, I, Fredrick, R.J., and Ramachandran, S. C., 1995. Impact of international harmonisation on biosafety regulations in biotechnology, p. 115-125. Proceeds. on biosafety regulations in biotechnology. Bangkok, Thailand.

Walsh, V., 1993. Demand, public markets and innovation in biotechnology. Science and Public Policy 25:138–156.

Williamson, M., 1991. Biocontrol risks, p. 394. Nature 353.

IPR Controversy and the Indian Seed Industry

Suri Sehgal
Proagro Seed Company Ltd., New Delhi, India

Introduction

The recent flurry of mergers, acquisitions, and alliances in the seed industry in North America and Europe can shed some light on the Intellectual Property Right (IPR) controversy in India regarding seeds. The events in North America and Europe have, to a large extent, been driven by IPR issues. India is not immune to these issues, especially since the introduction in 1993 of the Plant Variety Act (PVA) which led to an emotionally charged debate both within and outside the parliament. The following attempts to place the Indian experience of IPR in a broader global context and to draw several conclusions.

Success in the seed industry has usually been traced to the strength of a company's classical breeding program. Even the introduction of sophisticated molecular marker technology merely helped to optimize classical breeding, rather than fundamentally changing the competitive dynamics of the seed industry in the developed world. But with the advent of the first products resulting from genetic engineering, we are witnessing a significant restructuring of the seed industry in those countries. While it may be too soon to say what the restructured seed industry will look like, we can be sure that it will be heavily imprinted by existing intellectual property rights.

In this context the motto of the American Seed Trade Association, "First the Seed," has acquired a new twist of meaning. With farmers more and more likely to be planting genetically engineered seed, companies seeking to recover value from their technological developments are confronted by this fundamental truth of agricultural productivity long known to seedsmen. As with traditionally bred traits, seed is the carrier of genetically engineered traits, including insect and disease protection, herbicide tolerance, modified oil, and starch and protein content,to name a few. These technologies and their associated IPRs have thus led companies from such diverse businesses as agrochemicals, agricultural biotechnology, food processing, and many others, back to the seed.

Historical Perspective

For some time now the seed business outside of India and other developing countries has been considered a mature industry. The growth of the commercial seed industry was in fact due to the introduction of hybrids, especially hybrid corn in North America, hybrid sugarbeet in Europe, and hybrid vegetables in Southeast Asia. Of the $15 billion market in commercial seed at present, hybrids account for approximately 40% of sales and most of its profit.

Table 1. Herbicide-tolerant products

Year	Crop	Trait	Technology Supplier
1991	Corn	IMI*	American Cyanamid
1993	Soybeans	STS*	DuPont
1994	Cotton	BXN*	Rhone Poulenc
1995	Canola	Pursuit*	American Cyanamid
1995	Canola	Liberty Link™	AgrEvo
1996	Soybeans	Roundup Ready™	Monsanto
1996	Corn	Poast*	Univ. of Minnesota

IMI: Imidazolinone, STS: Sulfonyurea, BXN: Bromoxynil. Pursuit and Poast are brand names of American Cyanamid and BASF, respectively.
** Products selected by tissue culture or mutagenesis.*

In North America and Europe the hybrid seed industry grew from regionally based family businesses. The profitability of hybrids far outstripped that of nonhybrid open-pollinated seeds, leading to eventual consolidation in the industry and the dominance of several key companies in particular crops. The attraction of hybrids was obvious; when double cross corn hybrids were first commercialized in the U.S.

in early 1930s, they were priced at approximately 10–12 times the price of commercial grain. With the introduction of single crosses in the 1960s, hybrid corn seed prices jumped to 20–25 times the commodity price. In the 1970s these high margins attracted the attention of several agrochemical companies, that thought they saw in seed synergies with their own line of business. The acquisition of NK by Sandoz, of Funk Seeds by Ciba-Geigy, of Nickerson by Shell, and of Asgrow by Upjohn occurred during this period.

Table 2. Insect-protected products

Year	Crop	Trait	Technology Supplier
1996	Corn	Bt	Ciba-Geigy & Mycogen
1996	Cotton	Bt	Monsanto
1996	Potato	Bt	Monsanto
1998	Corn	Bt	Plant Genetic Systems
1998	Tomato	Bt	Plant Genetic Systems

Bt = Bacillus thuringiensis

In the 1980s agrochemical companies that were engaged in biotechnology research began to acquire seed companies with the realization that seed would be the primary delivery system for their new technologies, particularly biotechnology. They believed that delivering and capturing value from new input and output traits required control over the distribution channel. This brought companies such as Dupont, Elf Acquitaine (Sanofi), ICI, Monsanto, Rohm & Haas, and Unilever into the seed business. Their strategy was to capture margins along the length of the agribusiness chain from the laboratory to the field.

This strategy did not work for all the new entrants. First, the time required to convert the early new technologies into products took much longer than originally envisioned. Second, there was a conflict between the entrepreneurial management style of the comparatively smaller seed companies and the hierarchial style of most large chemical companies. Third, the learning curve has been longer and more complex than expected and has led to poor financial performance. Finally, unlike chemicals, seed cannot be marketed globally, but only in

Table 3. Selected transgenic products

Year	Crop	Trait	Technology Provider
1995/96	Tomato	Delayed Ripening	Calgene/Monsanto
1997	Corn	Bt	Monsanto, Dekalb, NK
1997	Canola (*B. napus*–spring)	SeedLink™	Plant Genetic Systems
1997	Corn	SeedLink™	Plant Genetic Systems
1997	Corn	Bt	Plant Genetic Systems
1997	Corn	Liberty Link	AgrEvo
1997	Cotton	Roundup Ready	Monsanto
1998	Mustard (*B. juncea*)	SeedLink™	Plant Genetic Systems
1998	Corn	Roundup Ready	Monsanto
1998	Soybeans	Liberty Link	AgrEvo
1998	Sugarbeet	Liberty Link	AgrEvo
1998+	Potato	Virus Protection	Monsanto
2000+	Canola (*B. rapa*)	SeedLink™	Plant Genetic Systems
2000+	Canola (*B. napus*–winter)	SeedLink™	Plant Genetic Systems
2000+	*B. oleracea*	SeedLink™	Plant Genetic Systems

agroclimatic regions similar to where it was developed. As a result, companies such as Shell, Pfizer, Rohm & Haas, Sanofi, Arco Chemical, Upjohn, and several others, began to divest their seed and biotech businesses in the 1990s.

Recent Developments

Lately, however, there has been a reassessment of the seed industry. With genetically engineered seed finally reaching the market, the benefits that plant biotech can bring to farmers is drawing attention. This realization is hastening the convergence of agricultural biotechnology and seed and chemical industries. This convergence is, in turn, driving change in the cost structure of the traditional seed business and in product pricing. We have, therefore, seen attempts to separate the value of technology from the value of the seed in the form of a "technology premium" to be paid by farmers when they purchase a product improved by biotechnology. For example, in 1966 the technology premium for Bt-based insect protection in cotton was over $30 per acre and roughly $10 per acre for corn.

In order to maximize value recovery, minimize threat of litigation, and secure access to technology, several strategic "partnerships" have been announced in recent months. Monsanto's acquisition of 49.9 % of Calgene, 45% of Dekalb and 100% of Agracetus were, to a large extent, driven by technology and IPR issues. Similarly, Empresas La Moderna's (ELM) acquisition of DNA Plant Technology was driven by the latter's technology and portfolio of delayed fruit-ripening patents. Dow Elanco's 46% participation in Mycogen was driven by the former's desire to secure access to Bt technology and other patents. On the other hand, the Zeneca and Vanderhave merger was triggered more by geographic fit and other considerations than by IPR considerations. The Ciba and Sandoz merger was also to a large extent driven by their mutual interests in chemical and pharmaceutical business. The most recent mergers and acquisitions include:

- April 1996—Monsanto acquires Agracetus
- March 1996—Ciba and Sandoz merge to create Novartis
- February 1996—Monsanto acquires 45% of Dekalb Genetics
- February 1996—Zeneca (ICI Seeds) and Vanderhave merge
- February 1996—ELM acquires DNA Plant Technology

- January 1996—Dow Elanco acquires 46% of Mycogen
- November 1995—Monsanto acquires 49.9% of Calgene

With the seed industry in a state of flux, new competitive strategies are expected to emerge. These strategies are likely to focus on four areas: (1) pricing based on separating the value of technology from the value of the seed; (2) market segmentation; (3) product development using classical breeding, genetic engineering, and technologies to reduce "cycle time," and (4) sales and distribution.

Table 4. Insect-protected plants

Subject	Components	Example	IPR
Plant variety	Germplasm	Protected variety	PVR
Selectable marker gene	Promoter	35S	Patent
	Coding sequence	*nptII*	Patent
Trait	Promoter	TR	Patent
	Coding sequence	*cry1Ab*	Patent
Transformation Technology	Ti-plasmid	pGV2260	Patent
Gene Expression Technology	Transcription Initiation	viral leader	Patent
	Translation Initiation	Joshi	—
	Codon usage	AT → CG	Patent
	Number of IPRs		8

In the flag ship U.S. corn seed market, for example, it is anticipated that the seed distribution system will undergo significant changes in the next five years to meet farmers' expanding needs for more sophisticated technology and information. There will be a growing trend towards multichannel distribution rather than the existing farmer-dealer system only. In hybrid seed corn, the very existence of many local and regional U.S. companies will depend upon the success of Holden's, Inc., a major supplier of foundation seed in integrating new technologies in its proprietary germplasm.

Genes and Enabling Technologies

These new technologies that are entering the market in the first products can be broadly divided into two major groups–genes and enabling technologies. Genes encode proteins, the major building blocks of all cells. For example, insect protection is due to the presence of proteins encoded by Bt genes, such as *cry1A(b), cry1A(c),* and *cry9C*. In 1996 roughly 1.8 million acres of Bt cotton, two hundred thousand acres of Bt corn, and eighteen thousand acres of Bt potato were planted in the U.S. Tolerance to the herbicide, Liberty™, is due to the presence of the genes *bar* and *pat*, while tolerance to the popular herbicide, Roundup™, is due to a gene encoding a mutant *EPSP* enzyme (Tables 1 and 2).

Table 5. Trait bundling opportunities

Crops	Input Traits	Output Traits
Corn	Bt, Roundup Ready, Liberty Link, SeedLink™	Oil, protein quality
Cotton	Bt, BXN, Roundup Ready, Liberty Link	Fiber quality
Soybeans	Roundup Ready, Liberty Link	Oil modification, protein quality
Canola	Liberty Link, Roundup Ready	Oil modification

In 1996 approximately two hundred thousand acres of Liberty™ tolerant canola and fifty thousand acres of Roundup™ tolerant canola were planted in Canada and approximately two million acres of Roundup™ tolerant soybeans in the U.S.

The important enabling technologies include plant transformation systems, selectable markers, gene expression techniques, and what can be called gene "silencing" technologies. Plant transformation is employed to insert specific genes into plant cells. Methods include using *Agrobacterium* as a vector (Ti mediated transformation), electroporation, or particle gun. The result of all such methods is that the plant cell incorporates novel DNA into its chromosomes. Since the incorporation of DNA (i.e., genes) is random, selectable markers such as *nptII* and *bar* are used to identify the transformed cells.

To be sure that the inserted genes function in their new environment, expression technology is employed in combination with specific gene promoters to specify the timing and location of gene expression. By contrast, gene silencing technologies, such as antisense, can be used to suppress gene expression. Calgene's "Flavr Saver" tomato uses antisense to suppress one of the genes responsible for ripening so that tomatoes can remain on the vine longer and become sweeter without going soft. (See Table 3 for a selection of transgenic products.)

IPR and "Freedom to Operate"

Even from these first products a rather complicated IPR pedigree emerges. At issue is the so-called "freedom to operate." Freedom to operate can be defined as legal access to all the technologies required to launch a product. This is particularly important because virtually all transgenic seeds either contain several technologies or are necessary for their development. Even in cases where a technology is novel and patented, it may be dependent on earlier developments and so cannot be freely used even by the inventor.

Already with single trait cases, the IPR issues are extremely complex. For example, a transgenic insect tolerant plant may involve Plant Variety Rights (PVR), plant patents, as well as several patents relating to transformation technology, the selectable marker employed, the gene coding for the insecticidal protein, the promoter, and various regulatory elements and modifications needed to adequately express genes in plant cells (Table 4).

Any IPR holder of even one element could block the commercialization of an insect tolerant variety based on this package of technologies. Alternative packages are as complicated. Because of the difficulties of sorting through various IPRs, the cost of doing seed business will increase, as will the likelihood of litigation. As companies seek to bundle traits to gain competitive advantage, gaining freedom to operate will become even more complicated. The permutations and combinations of various input and output traits give enormous opportunities to bundle and stack genes, provided one can resolve freedom to operate issues (Table 5).

IPR in India

India does not currently recognize IPRs in agriculture. In particular, Indian patent law excludes plants and animals from protection. However, since India became a signatory to the GATT (General Agreement on Tariffs and Trade), it is required to provide for the protection of microorganisms and microbiological processes. Additionally, plant varieties are to be protected either through patents or through some other *sui generis* system to be based on the International Plant Breeder's Rights Convention for the Protection of New Plant Varieties (the UPOV convention). The UPOV convention was signed in 1978 and revised in 1991. The India government accordingly drafted a Plant Variety Act (PVA) in 1993 that has cleared in the Lower House after considerable debate in the parliament but still awaits passage by the Upper House. The provisions of the PVA, below, are similar to those of the 1978 UPOV convention and its 1991 revisions—

> **Farmer's Privilege.** Farmers will be permitted to save seed of protected varieties for use on their own farm. They will also be permitted to exchange seed with other farmers. This "in kind" trade amongst farmers is common practice in India.
>
> **Essential Derived Varieties.** Essentially derived varieties cannot be sold without the permission of the original breeder. Such varieties are those that are derived from a protected source variety and that while clearly distinguishable, nevertheless retain all its essential characteristics. This provision is intended to protect breeders against the simple introduction into his variety of a distinguishing characteristic by mutation, repeated backcrossing, gene insertion, or otherwise without changing the overall genetics of the variety.
>
> **Breeder's Exemption.** This provision allows protected varieties to be freely used for further breeding.
>
> **Farmer's Rights.** In 1994 the Food and Agricultural Organization (FAO) requested countries to implement farmer's rights through the mechanism of the commission on Plant Genetic Resources. India was one of the first countries to comply and has included a Community and Farmer's Rights clause in the PVA.

In addition to the proposed PVA, India is also obliged by the GATT to introduce patents protection of microorganisms and microbiological processes within a grace period of 10 years.

It goes without saying that India will have to develop the appropriate administrative infrastructure to effectively manage a patent system. Patents are, of course, widely sought after in the developed world both by public institutes and private sector companies. Two basic justifications are advanced in support of patents–they encourage innovation through the disclosure of inventions and they promote research by granting exclusive access to the inventor. Several processes and products based on genetic engineering have already been patented in developed countries, including transformation technology, genes, vectors and such traits as insect protection, virus and stress tolerance, enhanced nutritional value, and pollination control. These technologies have immediate application in India. Transgenic seeds can reduce agricultural inputs and benefit India's resource-poor farmers. For this reason every effort should be made within the ten year grace period to accelerate the development of biotechnology and to convert existing technologies into products for the Indian farmer. If we fail to do so, then Indian farmers will be deprived of the very real fruits of this new technology. Additionally, the multinational corporations, that hold most of the biotechnology patents in developed countries, will come to dominate the IPR scene in India as well. There remain several outstanding questions regarding IPR in India and the PVA in particular. For example, how are plant breeders to be compensated? How is the PVA to be administered and by whom? What will its effect be on the seed industry? It should be noted in this regard that any costs incurred by seed companies due to farmer's rights or other IPR provisions will, in any case, be passed on to their customers. Since the farmer is the customer, he will be paying extra without directly benefitting. Since maintaining genetic diversity is good for everybody, asking farmers alone to bear the cost of it via seed price increases seems unfair. Concerning the three provisions of the PVA mentioned above, the following things can be observed:

- Although farmer's privilege has been criticized by the Seed Association of India, I believe that it is the right thing to do. For centuries India's farmers have played a critical role in conserving the land races that are the basis of commercial crop improvement programs. It is fair that they are not denied the reward of their past efforts. Any legislation which restricts farmers from saving seed for their own use or selling it to their neighbors should, therefore, be avoided. Restrictions on the sale of protected seed would unduly hamper the spread of new varieties. In this regard, it is pertinent to ask whether the Green Revolution would have been possible had the high yielding varieties of wheat and rice been protected in such a way as to stifle the transfer of seed between farmers.

- Because there appears to be a conflict between the provision for essentially derived varieties and farmer's privilege, which permits "in kind" trade of protected seed between farmers, the former should be deleted.

- Progress in plant breeding has been possible because of the free access to all germplasm. Restricting access to germplasm would adversely affect overall breeding progress and can lead to a decrease in genetic diversity. The breeder's exemption should therefore be maintained in the PVA. Since plant patents exclude the breeder's exemption, such patents should not be permitted.

Conclusions

Agricultural biotechnology can add significant value to India's agriculture, but it will not be possible for the public sector to develop and apply alternative or novel technologies on its own. Budget constraints and other factors, therefore, require India to encourage the private sector in this area. As I have indicated, the private sector will require an IPR regime similar to that outlined above if it is to play a dynamic role in developing the Indian seed industry and if it is to commit to the sizeable investment in research that this will require.✧

New Genes for Old Genes

B. S. Ahloowalia
Joint Food and Agriculture Organization, International Atomic Energy Agency Division, Vienna, Austria

Preamble

The processes of recombination, mutation, and selection have given rise to genetic diversity. Conscious modification of plant genomes began with the domestication of plants. With the rediscovery of the laws of heredity, the modifications of the genomes were placed on a scientific basis. It was another 50 years before we realized the molecular basis of plant genomes was in DNA. Until then no patents were issued on plant genomes or genes. Plant breeder's rights were first established in Europe in the mid-1960s, giving protection to specific genomes. With the emergence and refinement of the recombinant DNA technology, patenting of genes and their sequences has been allowed in some developed countries. The patent process has been extended from the genome to the gene. Many plant genes evolved and exist in the developing countries while the patents on these genes are being sought in developed countries. How do you justify a gene as the exclusive property of an individual or a corporation when it has existed in nature for millions of years? Who really owns these genes?

Since genes do not act on their own but only in concert with the rest of the genome, the concept of genome rights is more meaningful than that of the individual genes. The progress in plant breeding has so far depended heavily on the free flow and availability of germplasm. Most developing countries in Asia, Africa, and Latin America do not have plant breeder's rights or legal systems to enforce them. If gene patents are to be allowed, then who owns the following genes?

- Mutant genes from a wild species or land race or an unrelated species or genus that have been modified by conventional mutagenesis.

- Mutant genes from a wild species or land race or an unrelated species or genus that have been modified by conventional mutagenesis.
- Transgenes mutated either with mutagen or by sequence changes, insertion, or deletion of base pairs or relocation by transposons. And whose liability is a patented transgene transferred to the crop of a neighbor? Taking an old gene and claiming it as a new gene by producing its detailed structure can hardly justify its exclusive ownership.

Introduction

The molecules capable of self-replication evolved with the origin of life on earth. During the next 4.5 billion years, the process of multiplication, mutation, recombination, and selection gave rise to the present day-genetic diversity that *Homo sapiens*, as the dominant species, exploit and use to the best of our abilities for our survival and comfort. The plant genomes underwent evolution that mainly supported their survival, reproduction, and proliferation. With the domestication of plants about 10,000 years ago by neolithic humans, conscious selection speeded up the process of plant evolution and took a slightly different direction. As a result, the genomes of major crops such as wheat, barley, maize, rice, potato, beans, lentils, cassava, banana, sugarcane, and cotton underwent an accelerated change to meet the demand of the growing human population.

It has been estimated that if a 2-hour film was made to cover the history of the neolithic man, then the domestication of plants and animals would be shown only in the last 2 minutes and the period of modern technology from discovery of the steam engine onwards to the atomic age would be covered in only three seconds and that of recombinant DNA research less than 1/5 of a second.

With the rediscovery of the laws of heredity in 1900, the modification of plant genomes was placed on a scientific basis, but it took another 50 years before it was confirmed that the molecular basis of plant genomes was in DNA. During the next 2 decades, research on DNA showed that it was possible to isolate and clone genes and then reinsert them into the same or a different organism. With the emergence and refinement of this technology, gene isolation, cloning, sequencing, and reinsertion has become a routine procedure.

Plant genomes have been modified by recombining and shuffling the existing genes. The genes and their sequences have evolved for thousands of years and in many cases millions of years. The basic molecules that make up the genetic code, tRNA, are themselves very old (ancient molecules) and arose some 1 billion years ago with the earliest life forms of RNA. Hence, they underwent many mutations and changes during their evolution. Who owns these genes, genomes, and their ancestral genes?

Ownership of Genetic Diversity

The progress of plant breeding has so far depended heavily on free-flow and availability of germplasm and its unrestricted use. Plant genomes and specific genes have been freely distributed and used by the developed and developing countries alike without restrictions and costs. Much of the plant genetic diversity exists in the centers of diversity or centers of origin, which are mainly located in the economically developing countries of the world. The plant collections presently held in the Consultative Group for International Agricultural Research (CGIAR) centers have been declared as the property of all nations held in trust.

Why are patents on specific genes being sought and granted in some countries which have freely made use of germplasm resources from the worlds common gene pools? How can one justify the patenting of genes which become property of individuals or corporations even though the genes evolved and exist in an entirely different geographical area? Who really owns these genes? Moreover, a gene does not act on its own but only in concert with the rest of the genome. What would you do with a single gene if the appropriate genome does not exist? How can you justify the payment of royalty on a single gene when the remaining 99.99% of the genome is a product of evolution and is held in public trust? Hence, the concept of genome (variety) protection in which the gene is present would be more meaningful than that of the gene.

Plant Variety Protection

Before the mid-1960s, no ownership rights were claimed or issued on any seed propagated variety or a specific trait of a variety although under the Plant Patents Act of 1930 in the United States of America, it conveyed rights to an owner to exclude others from asexual propagation of a patented plant. However, a new mutation or a "sport" derived from a patented plant could be multiplied and sold without infringement of the original patent.

The concept of protecting varieties as the Plant Breeding Rights was established in the United Kingdom (UK) under the Plant Variety and Seeds Act in 1964. Plant patent systems were introduced in the rest of Europe in August 1968 under the Union of Plant Variety (UPOV) convention negotiated in December 1961 and in 1970 in the United States as the Plant Variety Protection Act. These acts provide protection (effectively a patent) to sexually propagated plants. In genetic terms, it has meant giving a kind of intellectual property protection on a plant genome. However, many developing countries in Asia, Africa, and Latin America even now do not have such plant breeder's rights or legal systems to enforce them.

Before Plant Breeder's Rights came into effect, the varieties produced by the breeders were available for multiplication, growing, and sale without restriction or payment of royalties to the breeders. Plant Breeder's Rights were set up to justify the recovery of the high costs of plant breeding and to compensate the plant breeding companies for their investment. As a result, any variety which is distinct, uniform and stable can be protected under this legislation. This has created a certain legal responsibility on the part of the breeders and seed multiplication companies to provide the genuine variety true to its properties to the growers.

In principle, such protected varieties can be used freely for breeding of new varieties, without restriction. However, plant variety protection rights delay introduction, evaluation and release of many new varieties in countries which do not have laws on variety protection because owner companies do not provide seed without agreements on potential royalties. Thus, there is no longer free and unrestricted flow of plant germplasm for breeding.

Patenting the Genes

By going a step further in some developed countries, the protection has now been extended to genes and gene sequences, and many of the processes and procedures involved in gene isolation, cloning, and gene transfer. The patent process, hence, has been extended from the genome (e.g., the plant breeder's rights) to the stretches of the molecules of the genome (the gene). This is likely to become a nightmare of the developing nations, which have neither laws nor institutions to contend with gene patents and the release of genetically modified organisms. A gene patent prevents the free use of the gene and the variety it is in without the consent and agreement of the patent holder. As we refine our understanding of the molecular world of matter at the atomic and subatomic level in physics, it is to be expected that the same shall follow for the depth of understanding of self-replicating molecules and their sequences. Would patents in the future be allowed on the atomic and subatomic rather than molecular structures of the genes?

If gene patents are to be allowed to own specific genes, then, who owns the following genes:

- Mutants of existing genes from wild species, land races, or unrelated species, which have been modified through conventional mutagenesis in

  Unsequenced state; i.e., known from phenotype only; e.g., dwarf; male sterility, restorer, monogerm, disease-resistance, or brown mid-rib.

  Sequenced state—the gene has been isolated, cloned, sequenced, mutated (say by irradiation) and reinserted into the same cultivar or another cultivar or species.

- Transgenes (genes transferred through recombinant DNA technology) but mutated with radiation or chemical mutagen.
- Transgenes mutated by sequence changes (e.g., insertion or deletion of base pairs) and reinserted by vector splicing or relocated in the genome by transposons, which may change their expression.

Genetic Liability

If a patented gene/transgene is transferred to a nontransgenic crop of a neighbor through natural crossing, whose liability is it? The ownership of the gene creates certain liabilities on the part of the owner in relation to the unwanted transfer of the gene. The liability dates back to the famous case of Rylands v. Fletcher and the ruling given in 1865 by Judge Blackburn in London. According to this ruling, if "you collect any material on your land and the material escaped and caused damage, then you are liable for the damage caused." Your liability extends even if you were not negligent in its escape. This applies to anything such as animals, filth, water, stench or for that reason to pieces of DNA. The transfer of such genes to the crops of the neighbor could lead to litigation problems since the technology allows precise detection of such genes by the same technology by which these have been isolated and cloned.

Conclusions

Patenting of genes and their sequences is being allowed in some developed countries, which is likely to become a nightmare of developing countries since such legislation does not exist and must be enacted in these countries. Patenting of genes and gene sequences instead of the genomes is neither wanted nor required and is full of pitfalls for the following reasons:

- Mutants of existing genes known either from their phenotype or from their sequenced state and reinserted as transgenes should not be allowed a patent since both natural and induced mutagenesis is a well known process.
- Transgenes mutated by sequence changes (insertions, deletions, relocation by transposons) and reinserted by vector splicing are new alleles of old genes. Hence, the mutant transgene does not become property of the patent holder of the transgene.
- Taking an old gene and claiming it as new by producing its detailed sequence and modification of its molecular structure cannot justify an exclusive ownership, and patents should not be awarded for such genes.
- Patenting of genes or their modified forms which have been present in the centers of origin for millions of years is equivalent to genetic piracy and theft and cannot be considered as an innovative process to merit a patent.
- Should the genes be patented, the patent holders must be held liable for the unwanted transfer of patented genes.✧

The views expressed in this paper are of the author and in no way should be taken as the reflection of the policy of the United Nations or that of its Agencies.

The IPR Debate for India

Swati Prakash
Harvard University, Cambridge, Massachusetts, USA

THE ISSUE OF intellectual property rights in biotechnology is a complex and volatile one. The concept of patenting life provokes an uncomfortable response in most of us. At the same time it is difficult not to get caught up in the excitement generated by the potential of biotechnology to break free of the traditional constraints of productivity and usher in a new era of sustainable agriculture. It is natural to want to promote innovation through protection. One of the biggest challengesfor both the scientists and the policymakers who are central to evolving a new system of intellectual property rights in this country to maintain a cultural, emotional, and political perspective on the matter. It is extremely important that they do not succumb to overemotionality on the one hand nor the false promise of Paradise through copying the Northern model of development on the other.

Any discussion of intellectual property rights in this country must be firmly placed within the context of India as a developing country whose R&D goals are significantly different from those of the industrialized world. I have heard many scientists and policy makers, primarily from the North, say again and again that the bottom line in research is money; that without assurance that a scientist will have exclusive rights over her innovations, she will have little incentive to produce. If it is indeed the case that biotechnological research will respond primarily to the pull of the market, then I have trouble understanding how biotechnology can be the key to lifting India's marginal farmers out of the grip of poverty. Those farmers who are most in need of agricultural biotechnology in this country are those who can least afford to pay royalties on seeds.

Just because much of the sophisticated technology developed in this country is imported from the industrialized countries of the North does not mean that India has to simultaneously import the mode of thinking of the North. Those who argue that innovation in science will not advance without fiscal incentives for individual scientists do a grave injustice to the rich history of community developed knowledge and innovation in this country. For example, many "traditional" or folk crops, including several relatives of major commercial crops, are notable for their adaptation to extreme environmental conditions such as drought or salinity. This resistance to stress is not just a result of natural selection and evolution; farmers have played a critical role in breeding these crops over the years to be able to withstand such harsh conditions. Do these farmers have property rights to their knowledge? If intellectual property rights are to be fair and consistent in this country then they should protect all plant breeders, and not just those who work in laboratories with expensive equipment. This is particularly important considering that the primary need for biotechnology in this country is to improve crops for the huge proportion of the rural population living in marginal environmental conditions.

Whereas it is obviously desirable to encourage innovation and creativity in plant biotechnology by assuring some form of intellectual property rights, we must balance this need with the need of the public for innovation in areas other than profit-generating ones. I would not naively argue that a system of intellectual property rights for agricultural products should not be implemented in this country; I understand only too well the impetus created by the globalization and privatization of the Indian economy and agree that it is important to promote innovation by providing incentives. However, I do hope that as India moves forward in the complicated process of developing a "sui generis" protection system, she remains conscious of the important differences between its economy and culture and those of the countries of the world with the dominant system of IPRs, and locates her final decision within the context of India's own needs and resources.✧

Symposium Participants

AMRITA N. ACHANTA, Tata Energy Research Institute, Lodi Road, New Delhi 110003, India, 4622241 VOX, *aachanta@teri.ernet.in*.

ABHA AGNIHOTRI, Tata Energy Research Institute, Habitat Place, Lodi Road, New Delhi, 110003, India, 4622246, 4601550 VOX, 91–11–4621770/4632609 FAX, *abhagni@teri.ernet.in*.

PRAMOD K. AGRAWAL, Proagro–PGS, A–311 Ansae Chambers, New Delhi 110003, India, 619–4185 VOX.

BEANT S. AHLOOWALIA, FAO/IAEA Division, Wagramer Strasse 5, P.O. Box 100, A–1400 Vienna, Austria, 43–1–2060, ext. 21623 VOX, 1–12645 Atom A TELEX, 43–1–20607 FAX, *ahloowal@ripol.iaea.or.at*.

GHAYUR ALAM, Center for Technology Studies, A 40 Hauz Kha 8, New Delhi 110016, India, 91–11–6965163 VOX, 91–11–6960947 FAX, *alam.cts@axcess.net.in*.

R. K. ANAND, I.A.R.I., Entomology Division, India, 91–11–5781482 VOX.

B. S. ATTRI, Director (Research), Ministry of Environment and Forests, Government of India, CGO Complex, Lodi Road, New Delhi 110003, India, 91–11–4360806 VOX, 91–11–4360806 FAX.

V. K. BATISH, NDRI (ICAR), Molecular Biology Laboratory, NDRI Karnal, India, 255467 VOX, 0189–250042 TELEX.

BERYL BENJERS, National Library of Medicine (NLM), 8600 Rockville Pike, Bethesda, MD 20892, USA, 301–496–3261 VOX, 301–402–2433 FAX, *bb78x@nih.gov*.

SABHYATA BHATIA, Tata Energy Research Institute, India Habitat Place, Lodhi Road, New Delhi 110003, India, 4601550, 4622246 VOX, 4621770 FAX, *sabhatia@teri.ernet.in*.

SACHIN CHATURVEDI, R.I.S. (MEA), Fourth Floor, India Habitat Center, Lodhi Road, New Delhi 110003, India, 91–11–4617709, 4617403 VOX.

KRISHNA P. S. CHAUCHAN, Ministry of Environment and Forests, Government of India, Paryavaran Bhavan, CGO Complex, Lodi Road, New Delhi 110003, India, 91–11–4360769 VOX, 91–11–4362746 FAX, *biokps@envfor.delhi.nic.in*.

ILA CHAUDHURI, Department of Botany, Calcutta University, 35 B. C. Road, Calcutta 700019, India, (033) 475–3681/2/3/4 VOX, (033) 241–3222 FAX, *rajat@cubmb.ernet.in*.

RAJAT CHAUDHURI, Department of Botany, Calcutta University, 35 B.C. Road, Calcutta 700019, India, 91–33–472–3681 VOX.

VIRENDER L. CHOPRA, Indian Agricultural Research Institute, Biotechnology Center, New Delhi 1100012, India, 91–11–5753713 VOX, 91–11–5753678 FAX, *guest@bic-iari@dbt.ernet.in*.

GORDON DOUGLAS CLEMENT, Queensland Pharmaceutical &, Research Institute Griffin University, Brisbane 04111, Australia, 61–7–382191366 VOX, 61–7–38491292 FAX, *g.clement@pri-qu.edu.au*.

PETER COUCHAT, Commissariat A L'energie Atomique, Direction Des Sciences du Virant, Department d' Ecoplysiologie, er de Microbiologie Vepetale, Centre de Cardarache, 13108 Saint Paul Lez Durance-Ceidex, France, 33–42–254656 FAX, 33–42–2527–53 VOX.

N. K. DADLANI, I.A.R.I., New Delhi, India, 91-11-5788768 VOX, 91-11-3384978 FAX.

DEBJANI DEY, Division of Entomology, I.A.R.I., New Delhi 110012, India, 5781482 VOX.

BISWAJIT DHAR, RIS for the Non-Aligned and Other Developing Countries, Zone IV Fourth Floor, India Habitat Center, Lodhi Road, New Delhi 110003, India, 91–11–4617136 VOX, 91–11–4628068 FAX, *panchmuk@giasdlol.usnl.net.in*.

SUNITA GARG, Wealth of India, NISCOM (CSIR), New Delhi, India, 91–11–5786201 VOX.

H. S. GAUR, Division of Nematology, I.A.R.I., New Delhi 110012, India, 91–11–5786626 VOX.

Y. D. GAUR, Division of Microbiology, I.A.R.I., New Delhi 110012, India, 91–11–5711905 VOX, 91–11–5766420 FAX, *guest@bic-iari.ren-nic.in*.

S. L. GUPTA, Agriculture Entomology, Division of Entomology, I.A.R.I., New Delhi 110012, India, 91–11–5781482 VOX.

SATISH C. GUPTA, Ballarpur Industries Ltd., 364, 16 Main, 4 T Block, Jayanagar, Bangalore 560041, India, 91–80–6637509 VOX, 91–80–6637372/6637421 FAX.

WOLFRAM HEMMER, Agency BATS, Clarastrasse 13, CH–4058 Basel, Switzerland, 41–61–6909314 VOX, 41–61–6909315 FAX, *hemmer@ubaclu.unbas.ch*.

S. M. ILYAS, Assistant Director General (IPR), I.C.A.R, Krishi Bhavan, New Delhi 110001, India, 91–11–3388991/234 VOX, 031–62249 TELEX, 91–11–3387293 FAX.

MITTUR JAGDISH, VMSRF, P.O. Box 406, KR Road, Bangalore 560004, India, (080) 661–1664 vox, (080) 661–2806 FAX.

BHARATH JAIRAJ, Centre for Environmental Law, WWF–India, 172–B, Lodi Estate, New Delhi 110003, India, 4616532, 4627586, 4693744 VOX, *cel@wwfind.ernet.in*.

NIRAJA G. JAYAL, Jawaharlal Nehru University, A–16/9, Vasant Vihar, New Delhi 110037, India, 91–11–6110988 VOX, 91–11–6872117 FAX.

R. C. JOSHI, Division of Agricultural Physics, I.A.R.I., New Delhi 110012, India, 91–11–5781128 VOX.

S. K. KAPOOR, Proagro Seed Co. Ltd., A–308 Ansal Chambers, New Delhi 110003, India.

NUTAN KAUSHIK, Tata Engergy Research Institute, Darbari Seth Block, India Habitat Place, Lodi Road, New Delhi–3, India, 4622246 VOX, 462 1770 FAX, *kaushikn@teri.ernet.in*.

B. K. KEAYLA, National Working Group on Patent Laws, A–388, Sarita Vihar, New Delhi 110044, India, 91–11–681–3311/694–7403 VOX, 91–11–694–7403 FAX.

EKRAMULLAH KHAN, I.A.R.I., Division of Nematology, New Delhi 110012, India, 91–11–5786626 VOX, 91–11–5740722 FAX.

JITENDRA P. KHURANA, Department Plant Molecular Biology, University Delhi South Campus, New Delhi 110021, India, 91–11–600669 VOX, 91–11–6886427 FAX, *bic-dusc@dbt.ernet.in.*

PARAMIJIT KHURANA, Department of Plant Molecular Biology, New Delhi 110021, India, University Delhi South Campus, 91–11–4671208 VOX.

MARTIN KOCHENDORFEI, BUKO Agro Coordination, Torshabe 7, 74542 Braunsbach, Germany, 0045/7506/275 VOX, 0045/7906/7497 FAX.

GITA KULSHRESTHA, Division of Ag. Extension, I.A.R.I., New Delhi 1100012, India, 91–11–5787390 VOX.

DINESH KUMAR, Div. of /Environmental Sciences, Indian Agricultural Research Institute, New Delhi 110012, India, 91–011–5781474/5730411 VOX, 91–011–5740722/5752066 FAX, 031–77161–IARI–IN TELEX, *guest@bic-iari-ren.mic.in.*

KAUSHAL KUMAR, Bureau of Indian Standards, Manak Bhavan, 9 Babadur Shah Zafar Marg, New Delhi 110002, India, 91–11–3230131, 8375, 9402 VOX, 91–11–3234062, 9399 FAX.

T. MAKESH KUMAR, Division of Plant Pathology, I.A.R.I., New Delhi 110012, India, 91–11–5781474 VOX.

LAKSHMI KUMARAN (ADVOCATE), B4/158 Safdarjung Enclave, New Delhi 110029, India, 91–11–619–2243, 2273, 2280 VOX, 91–11–619–7578, 616–1820 FAX, *lakshmi@giasdlol.usnl.net.in.*

ANIL KUMAR KUSH, Indo-American Hybrid Seeds, P. O. Box 7099, 17th Cross, 2nd 'A' Main, K.R. Road, BSK 2nd Stage, Bangalore 560070, India, 91–80–6650111 VOX, 91–80–6650479 FAX, *kush@iahs.sprint.smx.ems.vsnl.net.in.*

MALATHI LAKSHMIKUMARAN, Tata Energy Research Institute, Darbari Seth Block, Habitat Place, Lodhi Road, New Delhi 110003, India, 4622246, 4601550, ext. 250882514 VOX, 31–61593 Teri In TELEFAX, 91–11–4621770 or 4632609 FAX, *malaks@teri.ernet.in.*

BENITO JUANEZ MARG, New Delhi 110021, India.

VEDPAL S. MALIK, U.S. Department of Agriculture, Animal and Plant Health Inspection Service, Plant Protection and Quarantine, Scientific Services, 4700 River Road, Unit 133, Riverdale, MD 20737, USA, 301–734–6774 VOX, *vedpal.s.malik@usda.gov.*

PIOTR P. MARAROV, Russian Agricultural Attache, Chanakyapuri, New Delhi 110001, India, 91–11–611–06–42 VOX.

J. R. MATHUR, Rajasthan Agricultural University, Bikaner, Rajasthan, India, 91–11–98110–51418 VOX.

K. L. MEHRA, FAO, UNDP, Consultant, 38 Munirka Enclave, New Delhi 110067, India, 91–11–6178369/6108608 VOX.

S. L. MEHTA, Deputy Director-General (EDN), Indian Council of Agricultural Research, (Krishi Anusandhan Bhawan), Pusa, New Delhi 110012, India, 91–11–5747760 VOX, 91–11–5750932/5750676 FAX, *mehta@icar.ren.nic.in.*

GURUMURTI NATARAJAN, Greenthumb, 10, Second Gross Street, Dr. Radhakrishnan Nagar, Madras–600041, India, 91–44–990 0924 VOX, *gumuna@giasmd01.vsnl.net.in*.

N. K. PANDEY, Ballarpur Industries Ltd., 364, 16 Main, 4 T Block, Jayanagar, Bangalore 560041, India, 91–11–080–6637380 VOX, 91–11–080–6637372/6637421 FAX.

PANKAJ, Division of Nematology, I.A.R.I, New Delhi 110012, India, 91–11–5786626 VOX.

CHANDRA PRAKASH, 9CPS Singapore, 4282 B-V Vasantkunj, New Delhi 110070, India, 91–11–613–6359 VOX, 91–11–613–6359 FAX.

SWATI PRAKASH, Harvard University, School of Environmental Sciences, and Public Policy, 76 Dante Street, Somerville, MA 02143, USA, 617–666–6367 VOX, *prakosh@fors.harvard.edu*.

RAJESWARI S. RAINA, NISTADS (CSIR), Dr. K. S. Krishran Road, Pusa Campus, New Delhi 110012, India, 91–11–5729151 VOX, 81–31–77182–NSTD–IN TELEX, 91–11–5754640 FAX, *nistads@sirnetd.ernet.in*.

P. SRINIVASA RAO, Division of Genetics, Indian Agriculture Research Institute, New Delhi 110012, India.

S. VENKATA REDDY, Deputy Director, Ministry of Environment and Forestry, Paryarayan Bhawan, CGO Complex, Lodi Road, New Delhi 110003, India, 91–11–4361669 VOX.

SURINDER SEHGAL, Proagro Seed Company Limited, A–311, Ansal Chambers-1, 3, Bhikaji Cama Place, New Delhi 110012, India, 91–11–6885082 VOX, 91–11–6872084 FAX.

SIVRAMIAH SHANTHARAM, U.S. Department of Agriculture, Animal and Plant Health Inspection Service, Plant Protection and Quarantine, Scientific Services, 4700 River Road, Riverdale, MD 20737, USA, 301–734–4882, *shanthu.shantharam@usda.gov*.

B. SHARMA, Genetic Division, I.A.R.I., New Delhi 110012, India, 91–11–578–3077 VOX.

H. C. SHARMA, Floriculture & Horticulture Technology, I.A.R.I., New Delhi 110012, India, 91–11–5785214 VOX.

B. P. SINGH, Division of Agricultural Extension, Indian Agriculture Research Institute, Pusa, New Delhi 110012, India, 91–11–5781434 VOX, 91–11–5752006 FAX.

N. B. SINGH, ICAR, Krishi Bhawan, New Delhi 110003, India, 91–11–3388033 VOX.

RAM H. SINGH, Plant Pathology Division, Indian Agriculture Research Institute, New Delhi 110012, India, 91–11–5782428 VOX.

S. K. SINGH, Division of Fruit and Hort. Tech., I.A.R.I., New Delhi 110012, India.

S. K. SINGLA, NDRI (ICAR), Karnal, Haryana, India, 91–11–189–252819 VOX, 91–11–0189–252819 FAX.

ANIL SIROHI, Division of Nematology, I.A.R.I., New Delhi 110012, India, 91–11–5733834(R)/91–11– 5786626(O) VOX, *sirohi@iari.ernet.in*.

Y. N. S. SRIVASTAV, Division of Entomology, I.A.R.I., New Delhi 110012, India, 91–11–5781482 VOX.

B. G. SRIVASTAVA, Agricultural Entomology, Division of Entomology, I.A.R.I, New Delhi 110012, India, 91–11–5781482 VOX.

K. L. SRIVASTAVA, Agricultural Entomology, Division of Entomology, I.A.R.I, New Delhi 110012, India, 91–11–5781482 VOX.

MEENAKSHI SROVASTAVA, Division of Entomology, I.A.R.I., New Delhi 110012, India, 91–11–5783826 VOX.

RAJEEV K. UPADHYAY, Deputy Director (Plant Pathology), Directorate of Plant Protection Quarantine, Protection and Storage, NHIV, Faridabad–121001, India, 91–129–212149, 91–11–3385026 VOX, 91–11–3384182, 91–129–212125 FAX.

VATSALA, Wealth of India, National Institute for Science Communication, India, 91–11–5786301/233 VOX.

LINDA WORTHINGTON, Diversity Day, 4905 Delray Avenue, Suite 41, Bethesda, MD 20814, USA, 301–907–9350 VOX, *diversitymag@igc.apc.org*.

DOO SUCK YANG, Hanhyo Institutes of Technology, Sang, Daeya-Dong, Siheung-Shi Kyungki-do, Republic of Korea, 82–32–692–0276 VOX, 82–32–692–0416, 0420 FAX.

STEPHEN YARROW, Agriculture and Agri-Food Canada, 59 Camelot Court, Room 3012, Nepean, Ontario KIA OY9, Canada, 613–952–8000, ext. 4107 VOX, 613–941–9421 FAX, *syarrow@em.agr.ca*.

Index

C

D

E

F

O

P

R

S

T

U

V

W

Z